An Introduction to Natural Medicine

Natural Medicine

From Plant to Patient

Phytoversity

Published by **Phytoversity**

An Introduction to Natural Medicine: From Plant to Patient

ISBN-10: 1999648307
ISBN-13: 978-1999648305

First edition April 2018.

Published in the United Kingdom.

All images reproduced under a Creative Commons (CC) License unless otherwise indicated. Chemical structures reproduced from the ChemSpider database.

Disclaimer Any content provided herein is solely for educational and information purposes and in no way constitutes medical or clinical advice.

For my family.

Acknowledgments Special thanks are due to my wife and the following people, without whose help this book would not have come to fruition:

Dr Linda Anderson; Marianne Jennifer Datiles; Dr Rui Fang; Christina Harrison; Lucinda Hawksley; Prof. Michael Heinrich; Dr Yukari Ishikawa; Dr Christine Leon; Dr David J. Newman; Dr Thomas A.K. Prescott; Dr José Prieto Garcia; Dr Francesca Scotti; Prof. Monique Simmonds; the countless other inspirational people I have met along the way and last - but not least - my dear friend Rodney.

There are more things in heaven and earth, Horatio,

than are dreamt of in our philosophy.

- Hamlet (1.5.167-8), Hamlet to Horatio

Contents

Research

Analysis

Testing

Complementary and Alternative Medicine (CAM)

Traditional Medicine

Regulation

Case Studies

Preface

Go back just a few generations and your ancestors would likely have relied on medicines sourced directly from nature. In today's world of pre-packaged pills, the idea of foraging for plants to cure life-threatening illness may seem strange, yet - according to estimates - in some Western countries up to 80% of the population have, in their life time, used traditional medicines and, a similar proportion of the world's population still depend on them.

All medicines have their potential risks and benefits, but, in increasing numbers, we are drawn towards **natural medicines**. Our reasons are many; some profess a dislike of chemicals - yet it is impossible to find a 'chemical free' medicine; some find Western medicine impersonal; others just want to take safer drugs - but, used irresponsibly, a natural medicine can be just as dangerous as an 'unnatural' one.

However, there is an astonishing variety of potential remedies in nature's 'medicine cabinet' and I wanted to write a book which offered a short introduction to how natural medicines are discovered, produced and used.

There are still many conceptions as to what a natural medicine is - as evidenced by the many names for this field: homeopathy; herbal medicine; naturopathic medicine; complementary and alternative medicine, the list goes on... With the wealth of terms to choose from, I have settled on 'natural medicine' as I believe it best describes what this book will focus on: **medicines taken from nature**.

Natural medicines can come from microorganisms, land and sea animals, minerals, plants, fungi, even stones (see cover image); **in this book I will introduce natural medicines which come from plants.** I will consider the basic concept of medicines and how they work, give a short overview of the types of chemicals found in medicinal plants and then look at some of the techniques we have for turning these chemicals into medicines. I will look at the principles of medical systems such as *Ayurveda* and Traditional Chinese medicine (TCM) and, lastly, show examples of traditional remedies which have been turned into important pharmaceutical drugs.

This book also contains two important appendices, one: a list of **Herbal Medicines Granted a Traditional Herbal Registration** under the UK THR scheme and the other: **Guidance on Banned and Restricted Herbal Ingredients** - both from the UK regulator, the Medicines and Healthcare products Regulatory Agency (MHRA).

As readers may be unfamiliar with some of the terms used, I have highlighted these in red and offer further explanation in the Glossary of Terms (pp 177-182). **Key concepts** are highlighted in bold print. Please bear in mind that this book can only serve as a brief introduction; nevertheless I hope that you will find it a useful one.

Owen Durant

London

April 2018

Natural Medicine

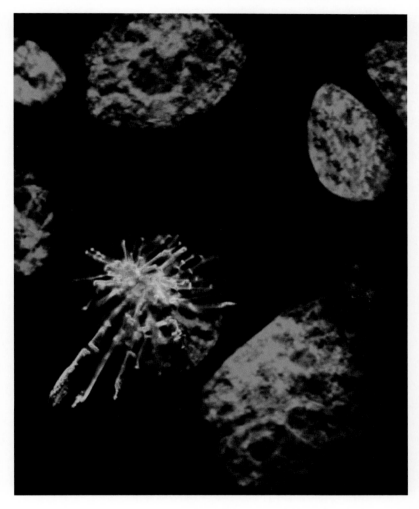

Prostate cancer cells (blue) being 'blasted' with curcumin (red/yellow) – a naturally occurring compound from *Curcuma longa* L., Light microscopy by Dr Khuloud T. Al-Jamal. (Khuloud T. Al-Jamal, Rebecca Klippstein & Izzat Suffian. CC BY 4.0)

What is a Medicine?

A **medicine** is, by the dictionary definition, *"a chemical substance used to treat, cure, prevent, or diagnose a disease or to promote well-being"*. In order to combat disease, medicines target the cells or pathways associated with specific conditions. For example: cancer drugs can destroy proliferating cancer cells and prevent their spreading (metastasis) through the body, while anti-inflammatory drugs can disrupt or block the pathways which signal the inflammatory response to infection or injury.

The UK regulator, the Medicines and Healthcare products Regulatory Agency (MHRA), outlines two main categories of medicine: **medicinal products**, which are substances, or combinations of substances, which *"may be used in or administered to human beings either with a view to restoring, correcting or modifying physiological functions by exerting a pharmacological, immunological or metabolic action"*, or *"to making a medical diagnosis"* and **medical devices** which *"do not achieve their principal intended action in or on the human body by pharmacological, immunological or metabolic means"*, but which may be *"assisted in their function by such means"*.

These definitions show the range of functions a medicine can perform, but there is also a third category of **borderline products** which do not fall neatly into the first two categories and these can be food supplements, biocides, cosmetic products or 'general products'.

But what are **herbal medicines**? According to the UK's Human Medicines Regulation (2012) a product is an herbal medicine if *"the active ingredients are herbal substances and or herbal preparations only"*. Herbal medicines exist in numerous forms, they can be: reduced or powdered; tinctures; extracts; essential oils; expressed juices or processed exudates.

Herbal preparations are defined by the MHRA as products produced when herbal substances are *"put through specific processes, such as: extraction; distillation; expression; fractionation; purification; concentration and fermentation"*.

A distillation furnace from Conrad Gessner's *The Newe Jewell of Health* (1576) (Wellcome Collection. CC BY 4.0)

Herbal preparations cannot make a health claim, but where prescribed for individual patients by herbalists - following a one-to-one consultation - they are currently unregulated. There have been unsuccessful legal attempts in the UK to prevent herbalists prescribing remedies in this way and to bring about sector-wide licensing.

 The EU operates a '**Register of authorised online sellers of medicines**' - requiring all online sellers of human medicines to display a country specific version of its **Common Logo** (inset) which links to a central register of authorized online sellers.

In the UK, an **herbal medicine with well-established use** must gain a license through an approved regulatory route, as with any other pharmaceutical medicine. **Traditional Herbal Medicines** can be approved through the **THR certification** scheme (see Regulation and Appendix 1). The 'pyramid' below summarizes the herbal medicines market, in the UK.

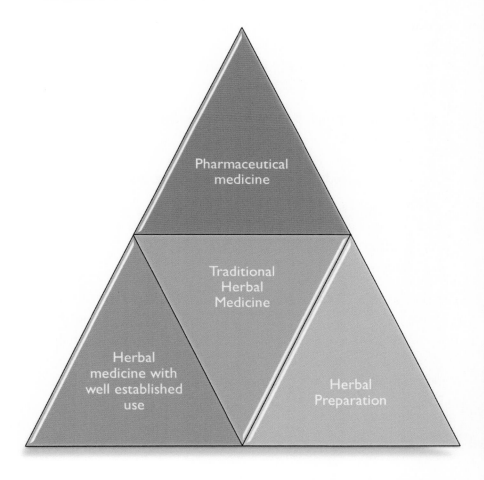

The darker green areas correspond with more tightly regulated categories of natural medicine and areas in lighter green with less tightly regulated categories of products.

What Makes a Medicine Natural?

All medicines induce a physiological response in the body which is, by definition, a natural organism. Where natural medicines differ from 'non-natural' drugs is not by their mode of action but by the **natural sources** from which they are derived.

We often hear about synthetic drugs, this does not automatically mean that these drugs are artificial; it means that they contain compounds which have been chemically reproduced or *synthesized*. So, in general terms, a drug can be:

- a **novel** compound unrelated to a naturally occurring compound;

- a synthesized version of **naturally occurring** compound;

- a **modified** version of a naturally occurring compound, or;

- a **mimic** of a naturally occurring compound (with similar structure, action, or effect).

The importance of **natural products** in **drug discovery** can be seen on the next page. Of the 1562 new drugs approved from 1981−2014, **50%** are **derived from natural products** (including modified and synthesized natural products and mimic drugs). If we consider anti-cancer drug approvals alone, this figure could be up to **75%**.

New Drugs Approved From 1981–2014

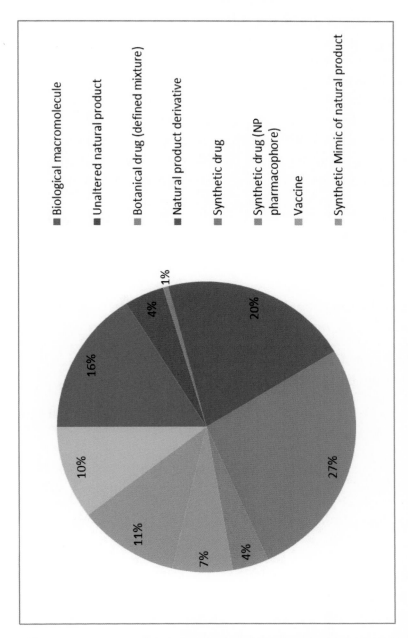

Data Source: Newman, D.J. & Cragg, G.M. (2016) Natural Products as Sources of New Drugs from 1981 to 2014. *J. Nat. Prod.* 2016, 79, 629–661

Why Modify Nature?

As well as potentially reducing side effects and improving efficacy, chemical modifications are commercially important as they may result in new drugs which offer the possibility of patents which can be extremely profitable for pharmaceutical companies. However, chemical modifications may also affect a compound's mode of action. To achieve better efficacy and fewer side effects from medicines, scientists need to have a detailed understanding of a medicine's behavior in the body, specifically: its **target**; its passage through the body to its target; and, its activity on reaching that target.

Natural Medicines in Action

The study of medicines in action is called **pharmacology**. To be effective any medicine must be absorbed, distributed, metabolized and excreted by the body. Pharmacologists refer to this sequence as **ADME** (absorption, distribution, metabolism and excretion):

Absorption – the entry of a medicine into the body through e.g. oral ingestion or direct injection into a tissue or an organ;

Distribution – the passage of a medicine to its intended target through the blood circulation;

Metabolism – the conversion of active compounds into therapeutic by-products (usually by cytochrome P450 enzymes in the liver);

Excretion – the passing of a medicine from the body in urine, feces or sweat.

Once compounds reach their targets their activity can take place. Some drugs have a very precise target, while others take a 'scatter gun' approach which can cause unintended side effects. Where a drug exerts activities involving various different targets we describe this as **polyvalent activity**. Depending on a drug's mechanism of action, its targets can be enzymes, proteins or DNA. Scientists can use techniques such as proteomics, genomics and metabolomics to look at the composition of targets and to find new ones for drugs.

We use the term pharmacodynamics to describe the ADME sequence from the drug's perspective – i.e. how the drug is transported through the body and, the term pharmacokinetics, to describe this sequence from the body's perspective – i.e. how the body responds to a drug's presence. Sometimes a 'T' for **Toxicity** is added – to form the acronym ADMET. Toxicity, in this context, refers to the damage that can potentially be caused to the body or its organs by a drug or its by-products.

Because of the way our bodies metabolize medicines, problems can arise when our metabolic 'machinery' is out of sync with the rate of processing required - it is possible for toxic compounds to rapidly build up in the body because processing is too slow or, for the drug to 'disappear' because processing is too fast.

Drug Interactions

Pharmaceutical drugs are largely single compounds, whereas the chemical profile of a medicinal plant usually consists of many compounds, acting together. Such complex interactions can cause side effects but may also offer benefits which can be lost in a single compound or modified version of a naturally occurring compound. Needless to say drug interactions are highly complex and difficult to predict, but generally, medicinal compounds may have one of the following effects on each other's activity:

Synergism – an increase the activity of another compound

Additivity – the combined actions prevent any measureable activity

Antagonism – a reversal or reduction of activity

Micro-particle drug delivery (Annie Cavanagh. CC BY-NC 4.0)

Another factor to consider is **drug dose** (posology) and pharmacologists also place great emphasis on measuring and achieving the appropriate **bioavailability** of a drug - i.e. its blood concentration and the quantity of a drug which reaches its target site.

An important point to bear in mind is that, all medicines, natural medicines included, can vary in effect from person to person. These variations may result from simple physical difference in the patient, for example weight or height, to genetic differences, lifestyle, allergic reactions or enzyme deficiencies. Consequently, the same care should be taken with prescribing and using a natural medicine, as with medicine from any other source.

Chapter Resources

MHRA Herbal and Homeopathic Medicines

https://gov.uk/topic/medicines-medical-devices-blood/herbal-homeopathic-medicines

NHS Herbal Medicines

https://nhs.uk/conditions/herbal-medicines/#potential-issues-with-herbal-medicines

Fake Medicines

https://fakemeds.campaign.gov.uk/

Register of Authorised Online Sellers of Medicines

https://medicine-seller-register.mhra.gov.uk/

Natural Medicine: Past to Present

Until the second half of the 20th century most people in the Western world used some form of natural medicine. With the advent of universal health care and the availability of affordable antibiotic drugs, many common infections were effectively eradicated and the popularity of natural medicines declined. Yet, large swathes of the world's population were untouched by this medical revolution - with up to 80% continuing to rely, in some form, on natural remedies handed down by millennia old traditions.

Where and when these healing traditions began is something of a mystery. Anthropologists suggest that - in response to their environment - early humans developed a 'sensory ecology' based on the appearance, smell, touch and taste of plants, through which they learned to find natural medicines. It has been suggested that people also learned about medicinal plants from studying the interaction between animals and plants.

In Ancient Egypt and China mythical 'bringers of knowledge' supposedly advanced popular understanding of how the natural world functioned. In Egypt the god Thoth and, in China, Shennong - the 'divine farmer' – were said to have delivered knowledge of medicine and agriculture which was passed down through oral legends.

This oral method of knowledge transmission began to change around 3,000 BC. Ancient Egyptians documented their medical knowledge into hieroglyphic texts in which lie the roots of Greek medicine and alchemy. Commonly misunderstood as the attempted manufacture of gold, alchemy was actually a sophisticated blend of astrology, sorcery and 'primitive' chemistry. The widespread practice of alchemy in natural medicine would continue until the scientific revolution of the 17th century.

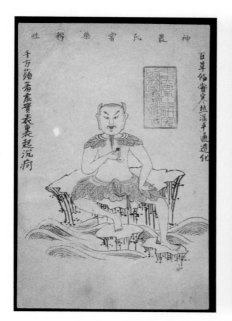

Top left: A portrait of Shennong (the divine farmer) from *Gudai yijia huaxiang* (Portraits of Ancient Physicians), created by Lin Zhong (C19 Chinese paintings of famous physician (Wellcome Collection. CC BY 4.0); Top right: Pale blue faience statuette of Thoth, Egyptian, 1500BC-100CB. Front 3/4 view of whole object on black background' (Science Museum, London. CC BY 4.0); Bottom: The Ebers papyrus c. 1500 BC - an Egyptian medical manuscript, named after Georg Ebers who purchased it at Luxor, Egypt in 1873-74

Hippocrates (460-370 BC) recapitulated some of the ancient concepts of medicine in his *Hippocratic Corpus* – which would go on to inspire Western medicine and ethics. His maxim "*let thy food be thy medicine and thy medicine be thy food*" is of continued relevance, even today - underscoring the inseparability of health and lifestyle. Less so, Hippocrates' widely believed humoral theory with its notion of the four inter-dependent humors: black bile, yellow bile, blood and phlegm.

Around 2,000 years ago in India, Ayurvedic healers consolidated their medical ideas into three books known as the Great Trilogy (the *Caraka Samhita, Sushruta Samhita,* and *Astanga Hridaya).* In China, between 200-250 AD, the *Shennong Bencaojing* (or *Pen-Ts'ao ching)* - a three volume work describing 365 herbal medicines attributed to the 'divine farmer' Shennong, classified plants according to their therapeutic properties and contained recognizable 'recipes' for producing medicines.

Dioscorides' *De Materia Medica* (50-70 AD) provided the first fundamentally Western pharmacopeia. On his travels in the Near and Middle East, the Greek physician Galen (129- 216) encountered a melting pot of medical traditions and cultures. *Unani* or Greco-Arab medicine, one such system to arise from this cultural exchange, was codified in Avicenna's the *Canon of Medicine (al-Qānūn fī aṭ-Ṭibb)* which - written in 1025 - remained in popular use until the 17th century.

In Europe, from the late 15th century onwards, new printing technology allowed for the reproduction and distribution of classic works such as the *Canon of Medicine* and *De Materia Medica,* as well as 'herbals' - elaborately illustrated books containing detailed information about the cultivation, production and folklore of herbal medicines.

Until the 16th century medicine was almost entirely plant or animal derived. A Swiss born physician, Paracelsus (1493-1541) began to devise 'chemical medicines' (e.g. mercury for the treatment of syphilis). He also insisted that medical texts previously published in Latin were translated into languages which 'common' people could understand.

L: A plate from *Hortus Sanitatus* 1484 – the first natural encyclopaedia depicting the mandrake root in 'human' form. (RBG Kew Library); R: The title page from a Latin translation of Avicenna's *Canon medicinae*. (Wellcome Collection. CC BY 4.0)

Paracelsus' ideas were, at first, viewed with suspicion - they contradicted classical humoral theory which had dictated the way in which illness was viewed and treatment chosen. There seemed to be no reason why medicine should re-invent its entire world view, but Paracelsus' ideas began to spark a quiet revolution - in their own way offering a roadmap from medieval ideas of humors and herbalism to modern concepts such as dose and drug design.

Across Europe medicinal plant gardens sprang up - first at Padua and Florence (1545) followed by Leiden (1590) and Montpellier (1593). In 1673 the Apothecaries' Company founded the Chelsea Physic Garden. In London and Paris scientific societies were formed - rarefied places where the great minds of the day would gather to exchange their findings and put forward bold new scientific theories.

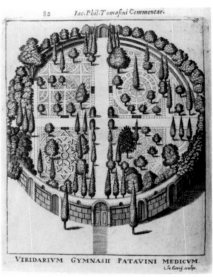

L: Aureolus Theophrastus Bombastus von Hohenheim *a.k.a.* Paracelsus. (reproduction, 1927, of an etching by Augustin Hirschvogel from 1538). R: Botanical garden at Padua. (Wellcome Collection. CC BY 4.0)

Nicholas Culpepper (1616-1654) an English physician published the *Complete Herbal* (1653) which became a best seller and has been in print ever since. Culpepper deliberately sold 'simples' at a much lower price than the controlling apothecaries and, like Paracelsus, a century earlier, was intent on popularizing knowledge of herbal medicines. Although Culpepper faced opposition from the medical establishment of his day, his accessible and systematic approach to medicine stimulated wide popular interest.

By 1668 the Merck pharmacy was founded in Germany, setting the scene for the mass production of medicines and the birth of a modern pharmaceutical industry. This drive towards modernity occurred against a backdrop of growing international trade between Europe, Asia and the 'new world' of the Americas. As a result of the British East India Company's explorations into Asia, exotic spices and medicines came into the hands of 'explorer-botanists', apothecaries, and herbalists. This trade in useful plants was accompanied by a growing interest in what we now call **ethnobotany** - the anthropological and socio-cultural study of the relationship between plants and the people who use them.

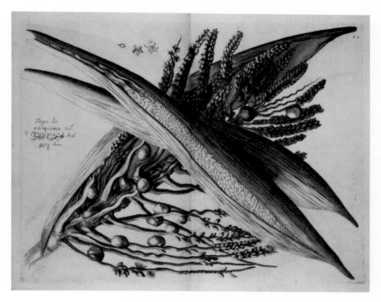

A plate from Hendrik van Rheede's *Hortus Indicus Malabaricus* (1678–1693) a survey of plants used in Ayurvedic medicine and an early example of ethnobotanical literature. (Wellcome Collection. CC BY 4.0).

Fresh advances in knowledge during the 18th century would lay the foundations for modern prescription medicines, which would ultimately usurp the place of medicinal plants. In 1796, Edward Jenner (1749-1823) discovered small pox could be prevented by administering a vaccine containing a tiny dose of small pox. Samuel

Hahnemman (1755-1843) by a similar principle, devised his 'law of similars' – giving rise to homeopathy. Johann Adam Schmidt (1759–1809) a German-Austrian surgeon would define the term pharmacognosy derived from the Greek words *pharmakon* (remedy) and *gignosco* (knowledge) to describe a growing body of research on medicinal plants.

Cover of *Floral poetry and the language of flowers* (1877)

In Victorian England the folk medicine tradition was still thriving, the so-called *'Language of Flowers*' (*floriography*) became a popular idea with its premise that flowers had innate personalities and emotions, harking back to the animistic world view of early healers, but in 'serious' circles herbalism was increasingly regarded as quaint and old-fashioned. Throughout the 19th century huge strides

28

were made in the understanding of plant chemistry. Dmitri Mendeleev (1834-1907) published the first periodic table in 1869; by 1900 Mikhail Tsvet (1872-1919) had devised chromatography - a technique which permitted the rapid separation and visualization of plant chemicals. Chemists now had tools to categorize medicinal compounds from plants and a growing understanding of their structures and properties.

The first pharmaceutical formulation of Aspirin launched by Bayer AG in 1899. (Bayer AG)

Meanwhile, Germany had become the major industrial producer of drugs in Europe; mass-producing plant-derived drugs such as heroin and morphine. Merck was a key player, with higher revenue than the major British pharmaceutical companies combined. Willow bark (*Salix alba* L.) had been recognised as a remedy for aches and fever since antiquity, its active principle salicin was derivatized and synthesized into acetylsalicylic acid by Bayer in 1897 and marketed as Aspirin in 1899. The accepted single compound paradigm of modern Western medicine had arrived.

World War II heralded the mass adoption of antibiotics, with the US shipping industrial quantities of Penicillin - isolated and purified from the mould *Penicillium* - to wounded soldiers on the battlefields of Europe. The post-war introduction of free universal health care (i.e. the NHS) was yet another blow to the popularity of herbal medicines. Although herbalism continued to be practiced, the incentive to use it was lessened by the availability of free, or heavily subsidized, prescription medicines.

Penicillium mould - the source of Penicillin.

In the post-war years sophisticated analytical techniques such as High Pressure Liquid Chromatography (HPLC) and Nuclear Magnetic Resonance (NMR) could, with just a small starting sample, identify not only the chemical constituents of a plant, but also the positions and conformations of its compounds. With access to this data, scientific interest in natural medicines increased and the National Cancer Institute (NCI) in the United States embarked upon what remains the largest mass screening of natural products in history - leading to the discovery of the blockbuster drugs Taxol®, Vincristine® and Vinblastine® for the treatment of cancers.

Developments in **synthetic chemistry**, too, meant that compounds could now be modified and synthesized on an even grander scale. Moreover, the discovery of patterns of chemical distribution in different plant families coupled with DNA profiling, allowed for the identification, purification and standardization of natural medicines in a way that had previously been impossible. But there were still risks as well as benefits to be gained from natural medicines. Following a spate of deaths resulting from the misidentification and mislabelling of traditional herbal medicines in the 1990s, a traditional herbal registration (THR) scheme was introduced in the UK to ensure the safety and quality of traditional herbal medicine products (THMPs).

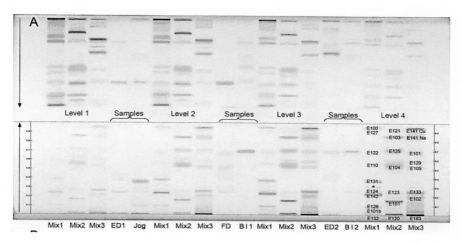

High-performance thin-layer chromatography (HPTLC) is a relatively inexpensive method for detecting adulteration or contamination in herbal mixtures. (G. Morlock, C. Oellig, CAMAG)

Today natural medicine is a global business, with the herbal supplements market worth an estimated $107 billion in 2017 and Traditional Chinese Medicine (TCM) $40 billion in China alone. The growing interest in natural medicines has brought with it ethical and moral concerns regarding the ownership of traditional ethnic knowledge (TEK) and the preservation of natural resources. Attempts to address these concerns have been made through global initiatives such as the Convention on Trade in Endangered Species; the Convention on Biodiversity and the Nagoya Protocol.

The debate around natural medicines continues, usually involving issues of safety and efficacy: are natural medicines safe and do they work? It is polarising debate with many rigidly sticking to their chosen system of healing. Can these approaches be reconciled?

Ethnopharmacology, the scientific study of the relationship between medicinal plants and the people who use them is perhaps the closest we have yet come to a unifying approach.

L: A selection of (possibly unlicensed) alternative medicines including Chinese, homeopathic and Bach flower remedies (Kate Whitley. CC BY 4.0). R: A traditional herbal registration (THR) product from the UK, showing the THR scheme certification mark.

One thing is for certain, with the growing popularity of natural medicines we owe it to ourselves to continue finding out more about the ways natural medicines can help or harm us; studying natural medicine and traditional knowledge in the light of hard scientific evidence can assist us in learning the lessons nature has to offer.

Categories of Natural Medicines

Traditionally, natural medicines have been classified by researchers according to a wide variety of intended uses – some of which are depicted below; the less commonly known terms are defined in the Glossary of Terms (pp 177-182).

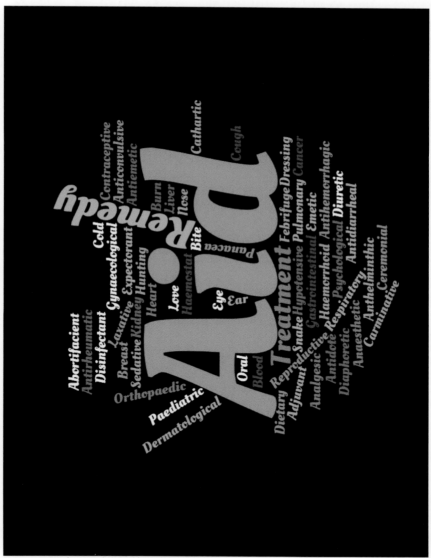

Image created using https://wordclouds.com terms via http://naeb.brit.org/

Food or Medicine?

One of the most fundamental questions one can ask about a plant with known health benefits is, *is it a food or a medicine?* With the rapid globalization of the natural products market, the question has become more pressing, and a slew of new terms – such as **botanicals, nutraceuticals** or **functional foods** - have been created to describe products which sit in the 'grey' area between nutritional and medicinal use.

Heinrich *et al.* (2018) suggest that rather than being discrete entities - foods and medicines exist in a kind of 'continuum' - spanning from staple foods, functional foods and food supplements to herbal remedies and prescription medicines:

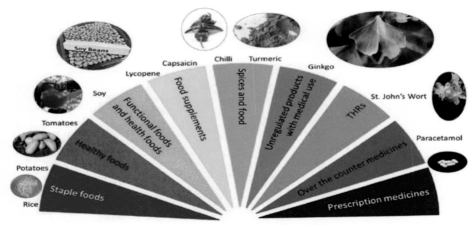

© 2018 Francesca Scotti & Michael Heinrich [unpublished draft] reproduced with the permission of the copyright holders.

The food or medicine debate is high on the agenda in the world of natural products, and has been intensified by the increasing body of scientific evidence regarding plants which have traditionally been consumed for health. For example, *Hibiscus sabdariffa* L. variously known as Roselle, Sorrel, Zoborodo, Sobolo, Bissap, Flore de Jamaica or Karkaday - depending on where in the world you are purchasing or consuming it - is variously used as a food, drink, flavoring and, in some clinical trials, has shown promise as an anti-hypertensive medicine.

Hibiscus sabdariffa L. – used as a food and a medicine (B. Simpson CC BY 3.0)

The debate as to what constitutes a food or medicine, is complicated not just by subjective opinion, but also by legislation - as there is no clear global definition of what is and what is not a food. Consequently, it is not unusual, to see herb extracts marketed as medicines in Europe while being sold as food supplements in another market, e.g. the United States.

The U.S. Food & Drug Administration (FDA) defines food supplements as: *'products taken by mouth that contain a dietary ingredient'* and goes on: *'these dietary ingredients may include vitamins, minerals, amino acids, and herbs or botanicals, as well as other substances that can be used to supplement the diet.'*

In the European Union (EU) 'food botanicals' are regulated by the European Food Safety Authority (EFSA). Health claims are prohibited on food botanicals, unless a dossier passes the EFSA's scientific evidence requirement which is more stringent than that which is applied to traditional herbal medicinal products (THMPs).

Botanicals - The term 'botanicals' has come to describe plant-derived products that are usually marketed as food supplements, although in the US another category of **botanical medicines** exists (see Regulation pp 105-107).

Functional Foods - A functional food can be defined as a food whose consumption extends beyond basic nourishment and is deemed to contribute to improved health.

Blueberries, a popular 'superfood' (Jim Clark CC BY 2.0)

Superfoods - Superfoods is a generic term for any food product with a substantial health claim. Common examples include: blueberries, *acai*, *chia* seeds and kale which are promoted, respectively, as sources of: antioxidants; omega 3; linoleic acid and vitamin K. Superfoods are a controversial area as the science behind food supplementation - such as the therapeutic value of antioxidants - is debatable. According to Harvard University's School of Public Health, studies "generally don't provide strong evidence that antioxidant supplements have a substantial impact on disease".

Nutraceuticals - A term which combines the words nutritional and pharmaceutical to lend scientific 'legitimacy' to a variety of supposedly functional foods.

Chapter Resources

Antioxidants

Harvard T.H. Chan School of Public Health

https://hsph.harvard.edu/nutritionsource/antioxidants/

Ethnobotany and Ethnopharmacology

Dr. Duke's Phytochemical and Ethnobotanical Databases

https://phytochem.nal.usda.gov/phytochem/search

The Journal of Ethnopharmacology

https://journals.elsevier.com/journal-of-ethnopharmacology

Historical Plant Use

Botanical.com

https://botanical.com/

Natural Products

Natural Products Global

http://naturalproductsglobal.com/

Plant Chemistry

Coffee plant (*Coffea Arabica* L.): flowering and fruiting stem. A popular source of caffeine. Watercolour. (Wellcome Collection. CC BY 4.0)

Atoms

The Periodic Table

Compounds

Classes of Medicinal Plant Compounds

Why Do Plants Make Medicinal Compounds?

Plant Chemistry

Plant chemistry, also known as **phytochemistry** (from the ancient Greek *phutón* or 'plant') is one of the richest areas of chemistry in terms of variety; with common chemical elements making up a seemingly infinite array of structures. Therefore, in introducing this topic, it might be helpful to recap some basic chemical concepts:

Atoms

The atom is a singular 'unit' of a chemical **element**. Each atom consists of a nucleus, containing protons and neutrons, around which electrons form an electrically charged outer 'shell'. The electrons on the outer shell govern an atom's attraction to, or repulsion from, other atoms.

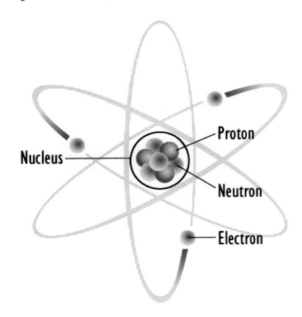

An atom (Fastfission CC BY-SA 3.0)

The number of electrons on the atom's outer shell also determine the atom's capacity to form bonds with other atoms – we call this capacity to form bonds an atom's **valence**

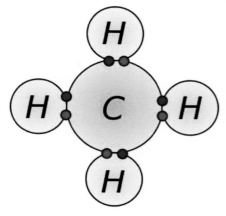

● Electron from hydrogen
● Electron from carbon

A methane molecule (DynaBlast - Created with Inkscape CC BY-SA 2.5)

The Periodic Table classifies elements – from left to right - by **atomic number** (the number of protons in each nucleus) in rows called **periods**.

The ordering of the periodic table from top to bottom provides indications about the chemical **group** an element belongs to and its properties, i.e. whether it is a solid, liquid, gas or metal.

In the periodic table on the next page, the elements most commonly found in plant compounds: hydrogen; carbon; nitrogen; oxygen and sulfur are highlighted.

The Periodic Table

1	2											13	14	15	16	17	18
hydrogen 1 **H** 1.0079																	helium 2 **He** 4.0026
lithium 3 **Li** 6.941	beryllium 4 **Be** 9.0122											boron 5 **B** 10.811	carbon 6 **C** 12.011	nitrogen 7 **N** 14.007	oxygen 8 **O** 15.999	fluorine 9 **F** 18.998	neon 10 **Ne** 20.180
sodium 11 **Na** 22.990	magnesium 12 **Mg** 24.305											aluminium 13 **Al** 26.982	silicon 14 **Si** 28.086	phosphorus 15 **P** 30.974	sulfur 16 **S** 32.065	chlorine 17 **Cl** 35.453	argon 18 **Ar** 39.948
potassium 19 **K** 39.098	calcium 20 **Ca** 40.078	scandium 21 **Sc** 44.956	titanium 22 **Ti** 47.867	vanadium 23 **V** 50.942	chromium 24 **Cr** 51.996	manganese 25 **Mn** 54.938	iron 26 **Fe** 55.845	cobalt 27 **Co** 58.933	nickel 28 **Ni** 58.693	copper 29 **Cu** 63.546	zinc 30 **Zn** 65.39	gallium 31 **Ga** 69.723	germanium 32 **Ge** 72.61	arsenic 33 **As** 74.922	selenium 34 **Se** 78.96	bromine 35 **Br** 79.904	krypton 36 **Kr** 83.80
rubidium 37 **Rb** 85.468	strontium 38 **Sr** 87.62	yttrium 39 **Y** 88.906	zirconium 40 **Zr** 91.224	niobium 41 **Nb** 92.906	molybdenum 42 **Mo** 95.94	technetium 43 **Tc** [98]	ruthenium 44 **Ru** 101.07	rhodium 45 **Rh** 102.91	palladium 46 **Pd** 106.42	silver 47 **Ag** 107.87	cadmium 48 **Cd** 112.41	indium 49 **In** 114.82	tin 50 **Sn** 118.71	antimony 51 **Sb** 121.76	tellurium 52 **Te** 127.60	iodine 53 **I** 126.90	xenon 54 **Xe** 131.29
caesium 55 **Cs** 132.91	barium 56 **Ba** 137.33	57-70 *	lutetium 71 **Lu** 174.97	hafnium 72 **Hf** 178.49	tantalum 73 **Ta** 180.95	tungsten 74 **W** 183.84	rhenium 75 **Re** 186.21	osmium 76 **Os** 190.23	iridium 77 **Ir** 192.22	platinum 78 **Pt** 195.08	gold 79 **Au** 196.97	mercury 80 **Hg** 200.59	thallium 81 **Tl** 204.38	lead 82 **Pb** 207.2	bismuth 83 **Bi** 208.98	polonium 84 **Po** [209]	astatine 85 **At** [210]	radon 86 **Rn** [222]
francium 87 **Fr** [223]	radium 88 **Ra** [226]	89-102 **	lawrencium 103 **Lr** [262]	rutherfordium 104 **Rf** [261]	dubnium 105 **Db** [262]	seaborgium 106 **Sg** [266]	bohrium 107 **Bh** [264]	hassium 108 **Hs** [269]	meitnerium 109 **Mt** [268]	ununnilium 110 **Uun** [271]	unununium 111 **Uuu** [272]	ununbium 112 **Uub** [277]		ununquadium 114 **Uuq** [289]			

lanthanum 57 **La** 138.91	cerium 58 **Ce** 140.12	praseodymium 59 **Pr** 140.91	neodymium 60 **Nd** 144.24	promethium 61 **Pm** [145]	samarium 62 **Sm** 150.36	europium 63 **Eu** 151.96	gadolinium 64 **Gd** 157.25	terbium 65 **Tb** 158.93	dysprosium 66 **Dy** 162.50	holmium 67 **Ho** 164.93	erbium 68 **Er** 167.26	thulium 69 **Tm** 168.93	ytterbium 70 **Yb** 173.04
actinium 89 **Ac** [227]	thorium 90 **Th** 232.04	protactinium 91 **Pa** 231.04	uranium 92 **U** 238.03	neptunium 93 **Np** [237]	plutonium 94 **Pu** [244]	americium 95 **Am** [243]	curium 96 **Cm** [247]	berkelium 97 **Bk** [247]	californium 98 **Cf** [251]	einsteinium 99 **Es** [252]	fermium 100 **Fm** [257]	mendelevium 101 **Md** [258]	nobelium 102 **No** [259]

Compounds

Most compounds (a combination of two or more elements bonded together) found in plants are composed from the following elements:

Element	Symbol	Atomic Weight (amu)	Valence
Carbon	C	12	4
Hydrogen	H	1	1
Oxygen	O	16	2
Nitrogen	N	14	3
Sulfur	S	32	6

If we imagine each **atom** of each **element** as a 'number' we can *add*, *subtract* or *multiply* these 'numbers' to produce the 'sum' of a **compound**. Our sum is expressed as a formula, e.g. caffeine is $C_8H_{10}N_4O_2$

A formula tells us the atomic composition of a compound – in the case of caffeine that it contains 8 carbons, 10 hydrogens, 4 nitrogens and 2 oxygens. This formula can be used to calculate a compound's total molecular weight (measured in atomic mass units or amu).

To work out the molecular weight of caffeine, we can perform the following calculation:

$C(12)x8+H(1)x10+N(14)x4+O(16)x2=194$.

Therefore, if we are trying to find **caffeine** in a mixture, we know its approximate mass is **194 amu**. This is very helpful when it comes to analyzing mixtures and determining their contents, as we shall see in the chapter on analysis.

Bonding

A bond is the connection (or linkage) between atoms in a compound. Bonds are necessary to form a compound and - to modify a compound - these bonds must be broken.

Hydrogen can form 1 bond with a neighboring atom, oxygen 2 bonds, nitrogen 3 bonds, carbon 4 bonds and sulfur, up to 6 bonds. The composition of bonds within a compound results in a net electrical charge (+ve, -ve or neutral) which determines the polarity of the compound.

Polarity

Polarity refers to the net electrical charge a molecule possesses. This value (measured as a dipole moment or μ) results from the sum of charges within a molecule's electron bonds. (e.g. bonds 1+-1=0 a neutral charge; 1+1=2 a positive charge: and -1+-1=-2 a negative charge.

For example, methane (CH_4) – shown on the following page - has a net neutral charge but, methanol (CH_3OH) – also shown - has a net positive charge.

With the addition of an oxygen atom, CH_4 forms an OH (hydroxy) group becoming CH_3OH - increasing its neutral charge so that it becomes positively charged or *polar*. (N.B. this is a theoretical reaction, methane cannot be directly converted into methanol, *yet*).

methane - CH_4 (neutral $\mu 0$)

methanol - CH_3OH (polar $\mu 1.7$)

The compound's name also changes from meth**ane** - with the suffix 'ane' representing the single bonds - to methan**ol** with the suffix 'ol' representing the addition of the hydroxy (OH) functional group. In chemistry naming, or *nomenclature*, is very important - as the name of a compound usually indicates the type of structure it is or the properties it has.

Classes of Medicinal Plant Compounds

A chemical class is a group of compounds which share key structural or functional characteristics. The suffix 'oid' means "like". Within an 'oid' (e.g. alkaloid, flavonoid or terpenoid) grouping, there can be many sub-types (e.g. monoterpenes, diterpenes or tetraterpenes). It is also important to remember that these descriptors are <u>not</u> mutually exclusive and a flavonoid, for example, could also be an alcohol (a flavonol). Three of the most widely observed chemical classes in plants are:

Alkaloids

Alkaloids contain nitrogen atoms and generally oxygen atoms too. They are often bitter in taste. Alkaloid compounds are well known as recreational drugs (e.g. caffeine and nicotine) and they frequently have psychoactive properties. Alkaloids also figure prominently as pharmaceutical drugs, e.g. quinine and morphine.

Alkaloids

caffeine $C_8H_{10}N_4O_2$ a powerful central nervous system stimulant and the main active compound in the Coffee plant:

nicotine, $C_{10}H_{14}N_2$ an addictive compound found in the Solanaceae (nightshade) family, which also functions as an insecticide.

quinine $C_{20}H_{24}N_2O_2$

morphine $C_{17}H_{19}NO_3$

Flavonoids

Flavonoids are based around phenolic ring structures, of the type shown below:

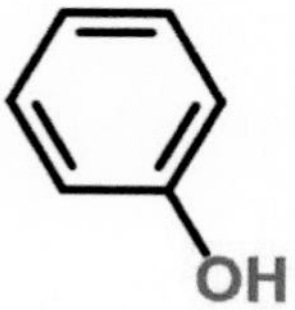

phenol - C_6H_6O

Flavonoids are found in most fruits and vegetables and are responsible for the pigmentation (coloring) of many plants. Although flavonoids often show antioxidant activity *in vitro*, there is little evidence of this activity *in vivo*. Quercetin, below left, is a flavonoid commonly found in multiple plant species; quercetin sulfate (right) containing a sulfur atom is a rarer compound.

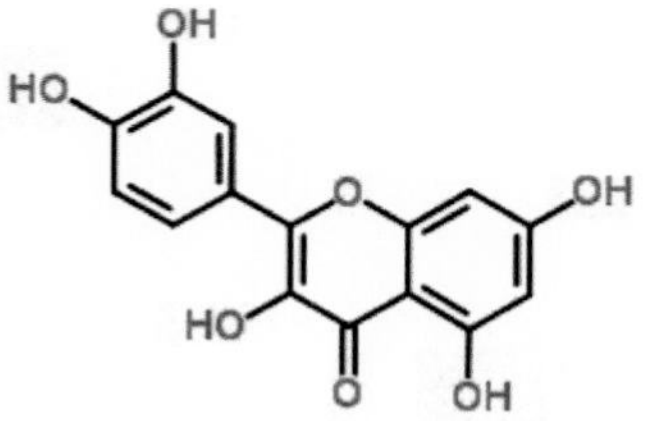

quercetin - $C_{15}H_{10}O_7$

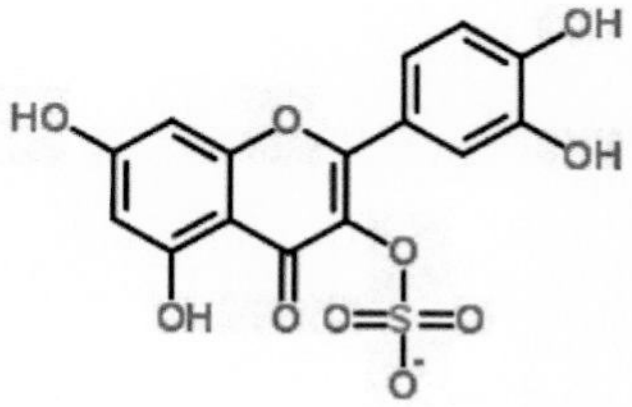

quercetin sulfate - $C_{15}H_9O_{10}S$

Terpenoids

Terpenoids are a diverse class of compounds which include essential oils or volatiles. Terpenoids are composed from 5-carbon isoprene sub-units as shown on the next page.

Terpenoids are categorized according to the numbers of these 5-carbon isoprene units that they contain.

Isoprene (C_5H_8)

Below are examples of monoterpene (10 carbon); diterpene (20 carbon) and tetraterpene (40 carbon) terpenoids:

monoterpene

limonene - $C_{10}H_{16}$

diterpene

marrubiin $C_{20}H_{28}O_4$

tetraterpene

β (beta) carotene - $C_{40}H_{56}$

Common Structures

Within these classes and, natural products generally, we see commonly repeating structures, e.g. aromatic rings, phenolic rings, benzene rings and **functional groups** such as: alcohols; amines; carboxylic acids; ketones and ethers. Functional groups are defined groups of atoms within a compound which contribute to its reaction properties and behavior. Often the letter 'R' is used to denote a generic organic (carbon) group attached to a compound.

benzene – C_6H_6

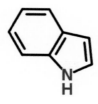

indole - C_8H_7N (an alkaloid containing an **aromatic** ring structure).

Stereochemistry

It is important to remember that stereochemistry or '3D chemistry' is a vital feature of chemical compounds. To put it simply, compounds do not exist merely as abstract drawings on a page – they can point, upwards, downwards, sideways and backwards giving them a quality called *dimensionality*.

For example, if we look at limonene - a monoterpene - we often find isomers - as below - which are identical in formula and molecular weight but structurally distinct, 'mirror images' of one another.

(S)-(-)-Limonene   (R)-(+)-Limonene

Such subtle differences are part and parcel of plant chemistry but they can have a major effect on the pharmacological behavior, detectability and, even, the smell of compounds. In the case of the limonene isomers pictured above, the (S)-(-)-limonene on the left has a citrus-like smell and the (R)-(+)-limonene on the right smells like pine trees.

Chemists use a variety of symbols - such as those pictured below - to represent these three-dimensional structures.

A '**broken wedge**' – is used to depict bonds below the plane (pointing away from viewer).

A '**dashed wedge**' – is used to depict bonds above the plane (pointing towards the viewer).

Why Do Plants Make Medicinal Compounds?

The majority of compounds from plants which are used medicinally are **secondary metabolites**, that is compounds which are not directly associated with growth and development but with environmental adaptation and survival, e.g. UV-protective flavonoids.

Many medicinal compounds are produced by plants to ward off predators and combat infection. These compounds' ability to combat disease against plant pathogens also give them the ability to fight against, frequently similar, human pathogens.

Medicinal plant compounds work by a number of means: by inhibiting growth of pathogens or by paralyzing or deterring predators. Defensive chemicals may also be produced in response to environmental stresses or, in the case of phytoalexins, in direct response to the presence of parasites. Plants also produce cyclotides which target multiple pathogens simultaneously.

Compounds of different classes each play a role in a plant's protective arsenal, with, for example, bitter tasting alkaloids and astringent tasting tannins playing a role in anti-herbivory. However, plant defenses are not solely chemical and it has been suggested that the more structural defenses a plant has (e.g. bristles or spikes) the less need it has for synthesizing these protective compounds. Some plant families, including ones with 'inhospitable' structures, e.g. the grasses (Poaceae), are rarely used medicinally and are instead primarily cultivated as foods.

Chemotaxonomy describes the association between chemical compounds and plant families and species. Some compounds appear in multiple families, whereas others are unique to a particular family or even a species of plant.

Chemotaxonomy can be applied to bio-prospecting, so that when we look for plants with potential medicinal activity we do so from a prior expectation of the types of compounds different plant families produce. Out of just over 400,000 known plant species, nearly 30,000 – or around 7.5% - have known medicinal usage.

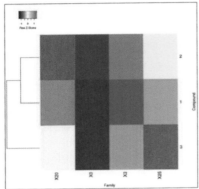

An example of a 'heat map' - showing the expression levels (highest concentrations in dark red, lowest in dark blue) of compounds in different plant families.

Chemotaxonomy can also be used to provide information which assists in the classification of plant species into genera and families (see next section). Chemotaxonomic data can be visualized in the form of 'heat maps' (above) or Principle Components Analysis (PCA) plots (below), which demonstrate the expression levels of different compounds in particular plant species or families.

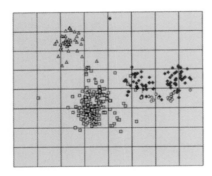

An example of a Principle Components Analysis (PCA) plot - showing the distribution of a compound in Family A; Family B; Family C; Family D.

Chapter Resources

Compounds

ChemSpider

http://chemspider.com/

Phytochemistry

Dr. Duke's Phytochemical and Ethnobotanical Databases

https://phytochem.nal.usda.gov/phytochem/search

Research

Artemisia annua L. (Sweet wormwood) the source of artemisinin - an anti-malarial drug.
(Sue Snell CC BY 4.0)

Identification and Naming

Cultivation and Harvesting

Preparation

Extraction

Fractionation

Purification

Research

Turning a plant into a standardized medicinal product is a complex process - summarized below. It involves the application of specialist techniques involving: **extraction**; **separation**; **analysis**; **purification**; **screening** and rigorous *in vitro* and *in vivo* **testing**. Very few active compounds from natural sources end up as licensed medicines.

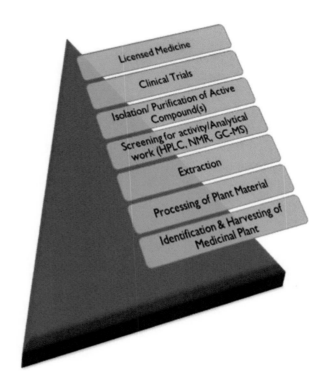

A simplified pathway from medicinal plant to licensed medicine.

Identification and Naming

The process of creating a natural medicine should begin by determining the species being used. Within a single plant family there are often many *genera* containing very similar species. Species form the lower level within the hierarchical system of natural classification, as marked in red below:

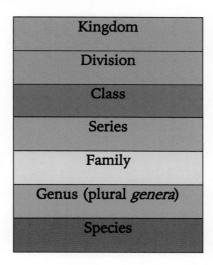

In order to create a standardized product it is crucial that we know that the plant selected is the species we think it is. We can use several resources to ensure that the plant harvested is correctly identified and to differentiate between species:

Botany - botanical experts who identify a plant species and assign a botanical (Latin) name.

DNA profiling/Whole Genome Sequencing - this gives us the genetic fingerprint of a plant species.

Ethnobotany – a discipline based on the interpretation of knowledge of plant use within different cultures or societies.

Herbarium/Voucher specimens – carefully chosen plant specimens with identifying information, such as botanical name, time and place of harvesting.

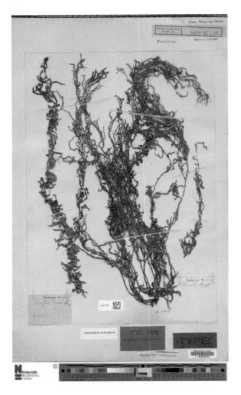

An example of a **type specimen** - used to represent a plant species: *Sargassum acinarium* L. Setch. (Source: Naturalis Biodiversity Center)

Microscopy - high powered lenses to visualize a plant's structure and find small but significant differences between species.

Systematics - the biological classification of plants according to their characteristics and relationships.

Taxonomy - the study, description and naming of variation between plants or other organisms.

Chemotaxonomy – see previous chapter.

Plant Names

One of the more challenging aspects of working in natural medicines is the interpretation of plant names. Plant names and naming systems can differ greatly according to culture and many species have multiple **common names**. To overcome these ambiguities researchers use **scientific names** – i.e. accepted Latin names from the Linnaean classification model (format: *Genus species*). Without the consistent application of these names there is always a risk that research and analysis may be based on the wrong plant species.

Cultivation and Harvesting

Cultivation can be summarized as the agricultural preparation of land - e.g. tilling, ploughing and irrigation. Voluntary guidelines are produced by the FDA and United States Department of Agriculture (USDA) to encourage Good Agricultural Practice (GAP) and Good Handling Practice (GHP). As well as stressing the importance of soil quality and climate - traditional systems often place great importance on the cultivation of medicinal plants in 'sacred' locations, time of planting and astrological factors such as moon phase.

Harvesting refers to the gathering of ripe crops (which can be carried out manually or mechanically). Depending on the season, time of day and place of harvesting, the chemical profile of a plant - its chemotype - can vary significantly. Even within the plant itself, different parts (e.g. leaf, root, flower or stem) will contain differing levels of particular chemicals.

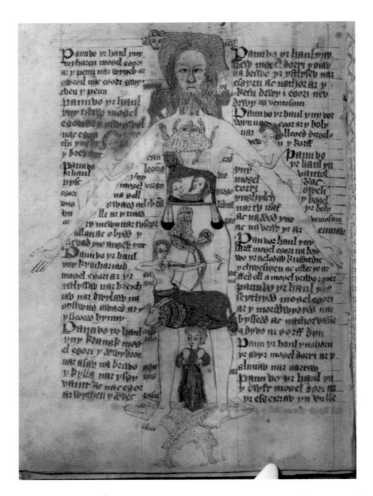

A 'zodiac man' from a 15th century Welsh medical manuscript; illustrating the role of astrology in traditional conceptions of human health.

Preparation

Once the correct plant and part has been identified and harvested, the process of preparation can begin. Careful preparation can help to remove contamination or 'noise' from a sample. A simple extraction may need very little preparation other than cutting and cleaning. For analytical or commercial purposes preparation is also likely to involve one or more of the following steps:

Air drying - exposure to warm air and ambient heat dries the sample and prevents sample deterioration.

Freeze drying - removes moisture from the sample at low temperature under a powerful vacuum.

Grinding - grinding into a fine powder with a purpose built mill or a mortar and pestle vastly increases the surface area of the sample, allowing more effective extraction to take place.

Extraction

Extraction is the single most important step in obtaining a plant medicine. As plants contain a 'cocktail' of chemicals, extraction requires solvents which help to isolate the desired compounds.

Extraction can be compared to 'apple bobbing' (see next page) - if we think of the apples as 'compounds', the children are the 'solvents' trying to 'grab' compounds from the mixture…

For anyone who has tried 'apple bobbing', it soon becomes obvious that getting the apples out is not easy. In extraction, we are also faced with the fact that frequently we are not sure exactly what 'apples' we are trying to extract. So, extraction strategies are often based on expectations of the potential compound(s) of interest, usually their chemical class and likely polarity.

Children 'apple bobbing' (Galia Pike)

An 'everyday extraction' of tea (*Camellia sinensis* L. Kuntze)
in a solvent of boiling water (H_2O).

Solvents

To extract compounds we actually use solvents which are liquids or gases (carriers). Solvents possess different polarities or chemical charges. A compound's polarity and other physical properties (e.g its pH) determine its affinity with the solvent being used.

To extract more polar compounds we use a more polar solvent and, for less polar compounds, a less polar solvent. Mid-polar solvents can be used to extract compounds which are neither very highly polar nor of low polarity. Below are some common solvents in increasing order of polarity from left to right; we can use them separately or, in combination, to obtain **fractions** containing compounds of different polarities.

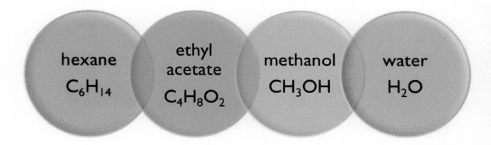

Large scale extractions are often performed using supercritical fluid extraction (SFE) in which compounds are extracted in a carrier gas - usually carbon dioxide (CO_2) - at elevated temperatures under raised atmospheric pressure. SFE is frequently used in food production, for example in the decaffeination of coffee, as it can extract large amounts of caffeine quickly and efficiently.

Fractionation

In nature compounds usually exist in a mixture. Fractionation separates a mixture into groups using different combinations of solvents. Once we have a fraction, it can be further divided into smaller and, progressively purer, sub-fractions by using different solvents at each step.

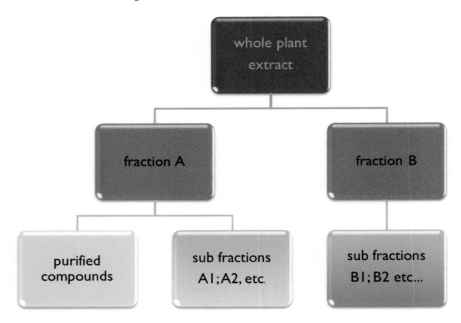

Purification

Purification is the process through which a pure compound is produced – as the name suggests, impurities within fractions are removed and a purified compound can then be confirmed using a range of analytical techniques. Purification can also be carried out by **preparative chromatography** – a process we will look at in the following chapter.

Chapter Resources

Plants

Plants of the World Online

http://powo.science.kew.org/

USDA Plants Database

https://plants.usda.gov/java/

Plant Names

The International Plant Names Index (IPNI)

http://ipni.org

Medicinal Plant Names Services (MPNS)

http://mpns.kew.org/

Analysis

A 700 MHz nuclear magnetic resonance (NMR) spectrometer. NMR is one of the most powerful tools available to analytical chemists.

Chromatography

Mass Spectrometry

Nuclear Magnetic Resonance

Hyphenated Techniques

Analysis

Each plant has a chemical 'fingerprint' - **analytical chemistry** is a tool with which we can interpret these 'fingerprints'. We may look at a mixture as a whole, or separate out its chemical constituents to look at compounds individually.

Analytical tools are, generally: **chromatographic** - where compounds are separated according to their affinity for either a solvent or for a solid material or, **spectroscopic** - where compounds are separated according to their atomic mass, charge and light absorbance. Chromatography and spectroscopy record information about chemical compounds using detectors which convert this information into spectra. Spectra, allow us to visualize information from a detector as peaks which correspond to the abundance of compounds in a mixture.

Spectroscopic methods also include **Nuclear Magnetic Resonance** (NMR) – which can identify the position of hydrogen and carbon atoms in a compound and provide information about a compound's stereochemistry. In recent years researchers have developed 'hyphenated' techniques, which combine chromatography and spectrometry to offer a more complete analysis.

Chromatography

Chromatography can be either **analytical** - to simply work out what is in a mixture - or **preparative** – in which compounds can be separated and collected into fractions for further purification. In many instances analysis and preparative work are carried out simultaneously.

Thin Layer Chromatography (TLC) – In TLC, compounds within a solvent (the mobile phase) bind at different positions on a layer of silica or paper (the stationary phase) where they separate into 'bands' according to their relative affinity to the mobile or stationary phase. These bands can be visualized, as below, either by the naked eye or under UV light.

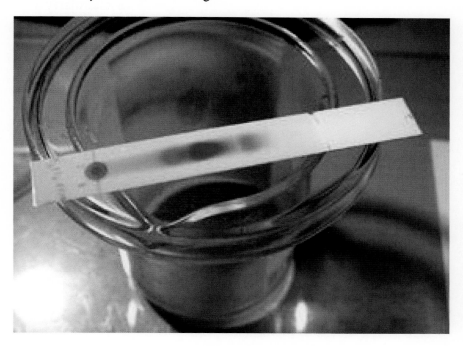

TLC of black ink in ethanol and water solvent (Natrij CC BY-SA 3.0)

High-performance Thin-Layer chromatography (HPTLC) - is used in industry for analysis and product standardization. HPTLC offers the speed and convenience of TLC but, due to the thinness (≤ 150 µm) and particle size (≤ 10 µm) of the stationary phase, it is capable of producing far better resolution than TLC.

High performance Liquid Chromatography (HPLC) – in an HPLC system (see next page) samples are injected in solvent at very high pressure through a narrow column. Compounds bind to the column for shorter or longer periods – depending on their relative polarity. The times taken for compounds to pass through the column are called **retention times**. These retention times are detected by a UV detector and 'translated' into a **chromatogram** on a computer attached to the HPLC equipment.

An HPLC System

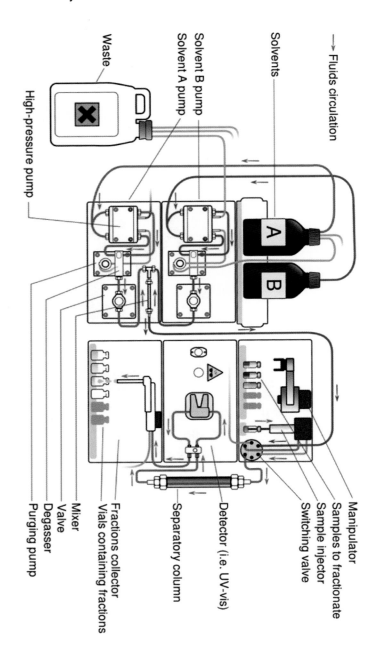

An HPLC system. (Yassine Mrabet CC BY 3.0)

An HPLC Chromatogram

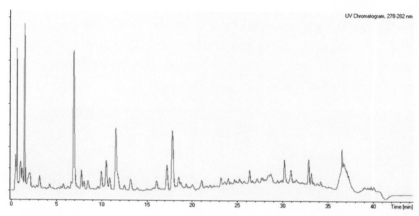

UV Chromatogram, 278-282 nm

An HPLC chromatogram showing results from an analysis of red wine. (Nono64 CC BY-SA 3.0)

There are various other types of chromatographic techniques; some involving only liquids (Liquid-Liquid Chromatography) and others which separate compounds, using a surface which deters either larger or smaller compounds from binding to it (Size Exclusion Chromatography).

Spectroscopy

Mass Spectrometry (MS) is one of the most widely used spectroscopic (or more precisely *spectrometric*) techniques. The process is summarized below.

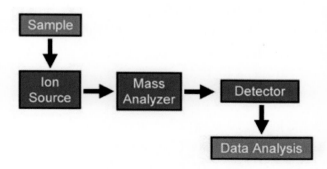

A diagram of the MS process (Kermit Murray CC BY-SA 3.0)

In a mass spectrometer a sample is bombarded with either positively charged or negatively charged ions causing it to fragment and the fragments are recorded by a detector, which calculates the fragments' mass/charge ratio and abundance and converts this data into spectra. As shown on the next page, the spectra reveal the relative intensity of components in a mixture.

An 'Orbitrap' mass spectrometer. (Nadina Wiórkiewicz CC BY 3.0)

In the example shown - caffeine appears at the far right of the spectra at 195 amu - having 'picked up' a hydrogen ion, adding 1 amu. The fragment at 138 amu with the highest peak is the most abundant and results from the caffeine molecule losing 2 carbons (24 amu), 3 hydrogens (3 amu), 1 nitrogen (14 amu) and 1 oxygen atom (16 amu).

At the extreme left (110 amu) the fragment is 84 amu smaller than caffeine - so we can infer that it has lost a fragment (marked by the red line around it) containing 3 carbons (36 amu), 3 hydrogens (3 amu), 1 nitrogen (14 amu) and 2 oxygens (32 amu).

Mass Spectra

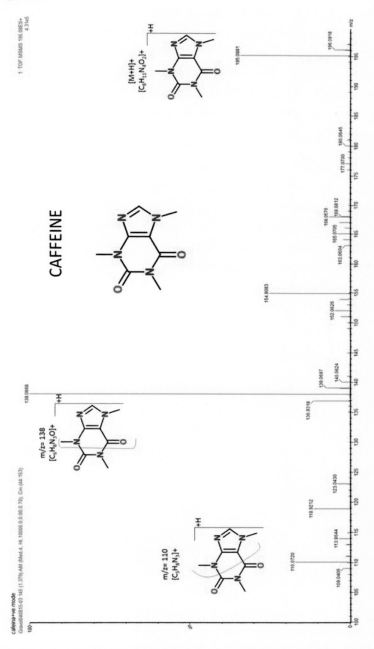

ESI+ spectra of caffeine (Santos Miranda Lopez: Antenor Orrego Private University, Trujillo, La Libertad, Peru)

Nuclear Magnetic Resonance (NMR) – when atoms are exposed to a magnetic field in an NMR device this 'excites' the atoms - causing them to spin. According to their positions atoms are **shielded** or **de-shielded** from the magnetic field and their degree of spinning differs accordingly. We call this degree of spinning **chemical shift**. In NMR, data produced from the detection of chemical shift in nuclei can be translated into a 'map' of atom positions in a compound.

NMR can detect only hydrogen ^{1}H; carbon ^{13}C or nitrogen ^{15}N isotopes – but, as carbon (and usually hydrogen) is present in all organic compounds, NMR can be an incredibly useful tool to add precision to an analysis. In the example below we see how chemical shift can be interpreted to determine hydrogen atoms' positions in different regions of a compound (ethanol - C_2H_5OH).

Ethanol

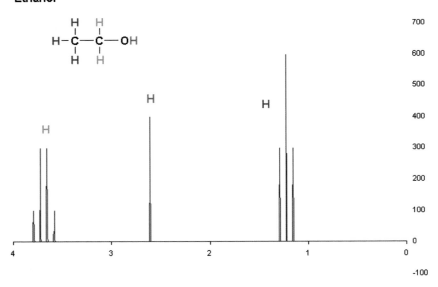

^{1}H-NMR spectrum of C_2H_5OH ethanol; the axis at the bottom represents chemical shift (measured in ppm from right to left) with the more exposed (de-shielded) regions showing a greater chemical shift. (T.Vanschaik CC BY-SA 3.0)

Hyphenated Techniques - 'hyphenated' techniques combine chromatography with spectrometry. Hyphenated techniques can be compared to High Dynamic Range (HDR) photography, in which different kinds of photographic exposures are superimposed to reveal a more detailed final image.

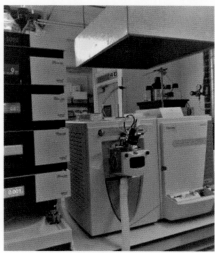

Top: an HDR exposure vs. a single exposure photo (Richard Huber CC BY-SA 3.0). Bottom: an LC-MS set-up (author).

Liquid Chromatography Mass Spectrometry LC-MS - combines HPLC with mass spectrometry. After passing through an HPLC column the sample is ionized and sprayed into a detector and its contents analyzed by the time they take to reach the detector.

Gas Chromatography - Mass Spectrometry GC-MS - uses gas (helium, nitrogen or hydrogen) as a carrier instead of a liquid solvent. The sample is injected into the gas and heated as it flows through a column and then into a detector which measures molecular mass.

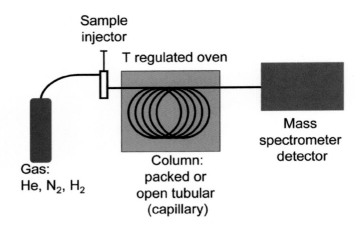

Schematic image of a Gas Chromatography – Mass Spectrometry (GC-MS) set up. (K. Murray. CC BY-SA 3.0)

Remember, data from all these methods can be combined and, because of the almost infinite variety of possible structures, often multiple data sources are needed to unambiguously identify compounds. The process of interpreting structural data is known as structure elucidation.

Chapter Resources

Analytical data

Spectral Database for Organic Compounds (SDBS)

http://sdbs.db.aist.go.jp/sdbs/cgi-bin/direct_frame_top.cgi

High Performance Liquid Chromatography

https://waters.com/waters/en_US/HPLC---High-Performance-Liquid-Chromatography-Beginner's-Guide/nav.htm?cid=10048919

Mass Spectrometry

http://waters.com/waters/en_GB/What-is-MS-and-How-does-it-Work%3F/nav.htm?cid=10073253&locale=en_GB

Testing

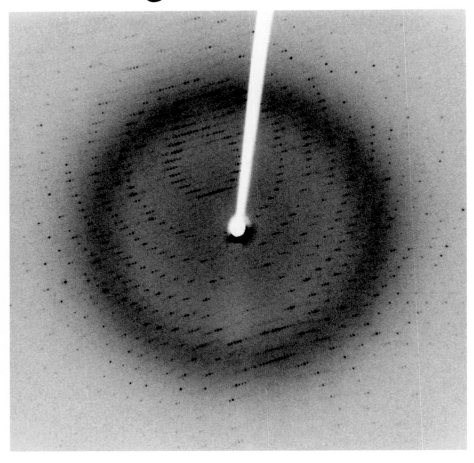

An x-ray diffraction (XRD) pattern. XRD detects the diffraction pattern of x-rays beamed through crystals of a powdered substance and can be used to detect impurities in drugs. (Jeff Dahl CC BY-SA 3.0)

Screening

Assays

Quality Control

Clinical Trials

After-Market Safety

Screening

Screening is a process through which **active compounds** and, therefore, potential **drug leads**, can be discovered. Screening tests a fraction or compound's efficacy against a selection of pre-defined **targets** - e.g. bacteria, enzymes or cancer cells. Screening may take place before, during or after analysis – depending on the approach being used.

Screening can be carried out in test tubes or micro plates in a method known as *in vitro* screening, or in living organisms, in a method referred to as *in vivo* screening. Screening identifies extracts with greater or lesser activity against targets and these results can be used to progressively guide the purification of fractions and isolation of compounds towards those which show more activity.

As the same compounds are often present in multiple species, they are frequently re-extracted – **de-replication** draws information from chemical databases of known active compounds - to prevent the same compounds from being repeatedly identified as active.

Assays

Assays are bio-chemical tests which test the target with fractions or isolated compounds in varying concentrations. The aim of assays in drug discovery is to find activity or **hits** and, if the compound has viability as a potential drug, it becomes a **lead**.

In order to save time, an automated technique called high-throughput screening (HTS) - in which thousands of compounds can be tested against a target simultaneously - is often used to find hits. It is also possible to carry out testing on a single plate of bacteria - where the size of the clear area around a compound or fraction (the zone of inhibition) demonstrates its activity against a bacterial strain.

Assays can also tell us the minimum amount of a substance needed to inhibit or kill a **pathogen**. This can be measured as an IC_{50} value - which tells us the concentration of a drug that is required for 50% inhibition *in vitro*. If we are looking for drugs which bind to specific targets to perform their activity, we use a comparable measure called EC_{50} - which tells us the concentration required to achieve 50% of this activity.

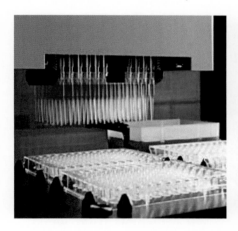 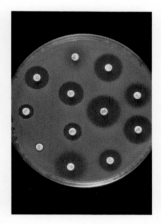

L: a high throughput screening (HTS) system (eupati.eu); R: a zone of inhibition assay.

Crucially, IC_{50} and EC_{50} values tell us if a drug is active in a concentration suitable for human consumption and, therefore, if it is a suitable lead.

If a suitable lead is identified, *in vitro* assays, such as those described, are followed by *in vivo* testing. In reality, a lead candidate must pass through many more hurdles to get to the market and only, an estimated 1 in 10,000, leads end up as pharmaceutical drugs.

Quality Control

Even after isolation, compounds can deteriorate or change (polymorphism) or become contaminated. A number of techniques can be used at this stage to establish the purity and stability of an isolated lead compound. Quality control measures (QC) prevent impurities and detect contamination and deterioration. Once QC measures have been applied and an Active Pharmaceutical Ingredient (API) has been produced, excipients are added so that the drug can be administered to participants in **clinical trials**.

Clinical Trials

Once proven effective in the lab, the process of bringing a drug to market involves a lengthy (sometimes >10 years) and costly (> US$1 billion) series of clinical trials:

Phase I trials assess the safety of a drug (several months) Phase I trials determine the effects of the drug on humans including its absorption, distribution metabolism, excretion and toxicity. Phase I trials also look at side effects as dosage levels are increased.

Phase II studies test the efficacy of a drug (several months - 2 years). Phase II trials usually involve randomized studies in which one group of patients receives the experimental drug, while a second 'control' group receives a standard treatment or placebo.

Phase III trials involve randomized and blind testing in 100s to 1,000s of patients. (several years). Phase III trials provide a drug developer with more data about possible adverse reactions.

Phase IV trials involve Post Marketing Surveillance Trials (ongoing) - after a drug has been approved for consumer sale. Phase IV trials compare the drug to others already on the market; monitor the drug's long-term effectiveness; patient impact and cost-effectiveness.

After Market Safety

Natural medicines, as with any drugs are subject to post-market surveillance - pharmacovigilance - and can be withdrawn from the market if toxicity or safety issues are identified.

Chapter Resources

Pharmaceutical Drugs

Medline Plus

https://medlineplus.gov/druginfo/herb_All.html

Drugs.com

https://drugs.com/drug_information.html

Complementary and alternative medicine

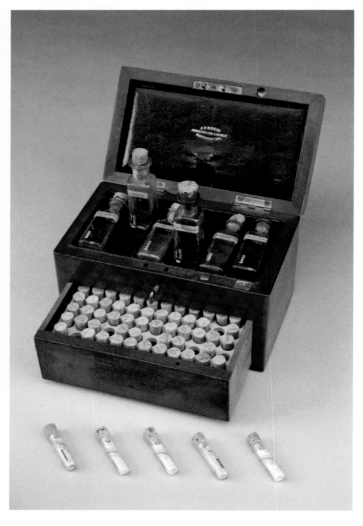

A homeopathic medicine chest, Northamptonshire, England, 1801. (Science Museum, London. CC BY 4.0)

Complementary and alternative medicine (CAM)

Integrative Medicine (IM)

Homeopathy

Naturopathic Medicine

Complementary and Alternative Medicine (CAM)

Complementary and alternative medicine (CAM) is a term covering a whole range of natural medicines and medical practices. **Complementary** refers to medicines which are sold to complement pharmaceutical products, whereas, **alternative** medicine is presented as an alternative option to 'conventional' treatment.

Many of these practices incorporate ideas and remedies from traditional medical systems, but because of the way that they are applied they do not themselves constitute a formal 'system'.

A woman and a man in the Lotus position (*padmasana*) illustrated with thermography
(Thermal Vision Research. CC BY 4.0)

CAM remedies are generally sold as: tinctures (alcoholic extracts); essential oils; fluid (e.g. water) extracts; capsules; tablets; ointments; gums; incense; herbs, roots or barks. CAM techniques can encompass anything from aromatherapy, yoga, reflexology and massage to cupping and magnet therapy.

Because of the range and scope of CAM the number of people who use CAM worldwide is huge; with around 40% of adults in the UK and North America having used at least one CAM remedy in the last year. The global CAM market is expected to reach $200 billion by 2025. These figures are backed up the growing profile of CAM in broadcast, print and social media.

A controversial issue with CAM is its use by patients with terminal conditions. A recent study in Europe, North America, Australia, and New Zealand found evidence that the use of CAM was increasing and that, almost half of cancer patients, had used some form of CAM (Trimble & Rajaraman 2017). There is concern in the medical community about how CAM can interact with cancer chemotherapy drugs and, in the U.S., cases have been reported of naturopathic practitioners prescribing non-approved CAM drugs to cancer patients.

Integrative Medicine (IM)

Arguably, integrative medicine, or integrated medicine (IM) as it is sometimes referred to, is simply complementary medicine by another name. However in IM a greater emphasis is placed on incorporating the latest scientific research.

The US Consortium of Academic Health Centers for Integrative Medicine – one of the global leaders in IM - states that IM aims *"to integrate biomedicine, the complexity of human beings, the intrinsic nature of healing and the rich diversity of therapeutic systems"* in order to *"transform healthcare"*.

In the UK, the Royal London Hospital for Integrated Medicine (RLHIM) defines IM as an approach which: *"brings together conventional medicine with safe and effective complementary medicine."* and that IM *"emphasizes the importance of the doctor-patient relationship and the use of all appropriate therapeutic approaches, healthcare professionals and disciplines to achieve healing and optimal health"*

Perhaps IM is a more palatable branding for CAM which has been labeled unscientific or pseudo-scientific by some. Maybe, with its greater emphasis on integrating conventional and traditional medicine, IM can widen the appeal of natural medicine among those who have historically been more skeptical about CAM usage.

Homeopathy

Homeopathy is often confused with herbal medicine. In fact they are very different things and homeopathy is an entire medical system with its own distinct medical philosophy. Homeopathy was the brainchild of Samuel Hahnemann (1755-1843) a German physician who went into medical practice at a time when blood-letting was still commonplace and many patients were dying needlessly from infection or botched surgical procedures.

Distressed by what he regarded as the barbarity of contemporary medical practice, Hahnemann began investigating 'safer' chemical medicines, conducting 'clinical trials' and developing his own system of homeopathy.

To differentiate homeopathy from other types of medicine Hahnemann coined the terms:

Allopathic - from the Greek *allos* ('other') which literally referred to medicines that were unrelated or 'other' to the cause of a disease or its symptoms – as Hahnemman viewed the commonplace medicine of his time.

Antipathic - from the Greek *anti* ('opposite' or 'against') which referred to medicines based on the theory *'opposites cure opposites'* which had been a popular concept from the time of Galen.

Hahnemann's own **Homeopathic** system - from the Greek *homeo* ('similar') - was based on the idea that *'like cures like'.* In Hahnneman's eyes, this meant that agents which were capable of causing the symptoms of a disease could also cure that disease.

Today, many homeopaths advocate that the smaller the dose is, the more powerful it becomes and they call this phenomenon **potentization**.

 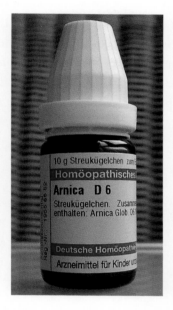

L: A portrait of Dr. Samuel Hahnemann. Founder of Homeopathy (Wellcome Collection CC BY 4.0) R: A modern homeopathic preparation, containing *Arnica montana* L. at a dilution of one part in a million (10^{-6}). (Abalg CC BY-SA 3.0)

Homeopaths claim potentization can be explained by water 'memory' which allows a compound to be 'imprinted' in a solution. **Succussion** or vigorous shaking unlocks the 'vital energy' within the medicine.

These ideas runs counter to the observations of modern science which suggest that a compound becomes undetectable below the molar limit (1 part in 1×10^{24}) and hence cannot be bioavailable or capable of generating any therapeutic effect. Many regard homeopathy as a placebo medicine, although advocates of homeopathy suggest that its apparent efficacy on animals demonstrates that homeopathy must be capable of producing a therapeutic effect.

Naturopathic Medicine

Naturopathic medicine, or naturopathy, has come to describe a therapeutic approach which emphasizes the importance of **self-healing**. In naturopathy, the fundamentals of good health can be achieved by regulating lifestyle. Depending on the preferences of the practitioner and patient, naturopathic medicine can encompass dietetics, botanical medicine, psychotherapy, naturopathic manipulative therapy, minor surgery, prescription medications, naturopathic obstetrics (natural childbirth), homeopathy and acupuncture.

A tin of organic powdered *Coriandrum sativum* L. - also known as cilantro, coriander or Chinese parsley – with a sprig of *Rosmarinus officinalis* L. (rosemary) in the foreground.

20 US states and three US territories have licensing or registration laws for Naturopathic Doctors. Naturopathic medicine is widely practised in India, where it is state-regulated, by a government Department of Ayurveda, Yoga and Naturopathy, Unani, Siddha and Homoeopathy (AYUSH).

Chapter Resources

Homeopathy

Homéopathe International

http://homeoint.org/english/index.htm

Integrative Medicine

US Consortium of Academic Health Centers for Integrative Medicine

https://imconsortium.org/

The American Association of Naturopathic Physicians

Naturopathic Medicine

https://naturopathic.org/

Traditional
Medicine

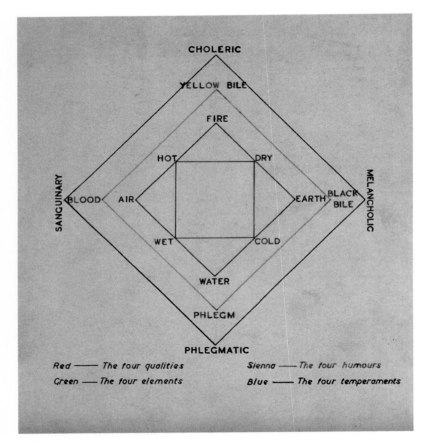

'Four' showing in Red, the four qualities; in green, the four elements; in sienna, the four humors and, in blue, the four temperaments.   (Wellcome Collection. CC BY 4.0)

Systems of Traditional Medicine

Traditional Chinese Medicine (TCM)

Ayurveda

Unani

Systems of Traditional Medicine

Two of the most widely known and still practiced schools of traditional medicine are *Ayurveda* and Traditional Chinese Medicine (TCM). Both Ayurveda and TCM are centered on the idea that an all pervasive 'life force' (in Ayurveda *prana* and in TCM *qi*) is connected to the human body. When this *prana* or *qi* is interrupted or unbalanced our bodies become sick.

In both Ayurveda and TCM our connection to the 'life force' can be optimized by consuming foods and medicines which maintain a 'balanced' state. Both systems emphasize the importance of balance in our lifestyle and habits. Diet, meditation and exercise (e.g. *Yoga* and *Tai-chi*) can play a part in sustaining a balanced state. Eastern life philosophy is part and parcel of these medical systems and by simply isolating compounds from traditional remedies it is impossible to grasp their holistic quality.

Traditional medical systems saw a resurgence in interest in the 19th and 20th centuries and, as a result of sustained contact with western biomedicine, have evolved and re-shaped to meet the changing demands and expectations of consumers.

Migration has also played a vital role in the dissemination of traditional medical knowledge across the globe. This dissemination of knowledge has worked to keep medical traditions alive and to create new ones.

For example, migration took Chinese medicine into Japan from the 6th century onwards – where it was re-imagined – and, is known to this day, as *Kampō* medicine. Migration has also facilitated the spread of traditional medicine from South and Central America and the Caribbean and numerous other regional traditions into diaspora communities worldwide.

Traditional Chinese Medicine (TCM)

Many readers will be familiar with the Chinese *yin* and *yang* symbol below. The *yin* represents the passive (female) state and the *yang* its active (male) counterpart:

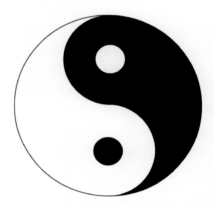

Rather than opposing one another, *yin* and *yang* - in a state of perfect balance - form a beautiful and harmonious whole. It is with the aim of achieving this harmonious state that different TCM medicines or medicinal mixtures are prescribed.

Qi (pronounced 'chee') – according to TCM philosophy - circulates through the universe and the body via a system of channels or meridians. TCM remedies and techniques such as acupuncture, cupping, moxibustion and *tu ina* massage can prevent disease and restore health by unblocking these channels to ensure the healthy circulation of *qi*. Correspondences are drawn between the five putative elements: wood; fire; earth; metal and water and specific organs and body systems - e.g. between fire and the heart and blood vessels - and treatments are chosen to balance these elements.

A TCM remedy can contain up to twenty herbs (although more typically this may be only 5 - fifteen) making it very difficult to evaluate the synergism of its ingredients. The herb components in a mixture are given names such as 'emperor', 'prime minister',

'minister' and 'envoy' – based on their role in this synergy. TCM remedies are produced by an elaborate array of processes including: brewing; stir frying; steaming; stewing and, even, fermentation into medicinal wines.

TCM, as we know it today, differs markedly from the disparate historical medical practices of China. In the same way that the traditional Chinese language was standardized and simplified in the 20th century, in post-revolutionary China, many of Chinese medicine's regional and cultural differences were discarded to create a 'universal system' under the umbrella of TCM. In the 1960s, so called 'barefoot doctors', often rural farmers with limited training, were recruited *en masse* to diagnose illness and prescribe TCM medication to villagers in remote regions across China.

A 'barefoot doctor' attends to a child. (Source unknown)

The state advocacy of TCM practice has led to a situation where TCM has an enormous take-up and is now estimated by analysts to constitute a third of sales in China's $117bn pharmaceutical market. In July 2017 a law was enacted by the Chinese government granting TCM equal status with Western medicine. It is now not uncommon to see TCM prescribed in state hospitals, where traditional remedies are prescribed, among other things, as adjuvants to conventional chemotherapy.

However, there are still major issues around regulating the quality of TCM drugs exported on to the international market, where low-quality or mislabeled products can end up in the hands of unsuspecting consumers. The Good Practice in Traditional Chinese Medicine Research Association (GP-TCM) initiative - funded by the European Commission - has spearheaded efforts to promote good practice in TCM research and development.

A Chinese Medicine shop in Chinatown, London – one of many established by Chinese communities across the world. (Jordiferrer CC-BY-SA 4.0)

In the UK, self-regulatory bodies such as the Register of Chinese Herbal Medicine (RCHM) and the Association of Traditional Chinese Medicine and Acupuncture (ATCM) uphold codes of conduct and ethics for participating TCM practitioners.

Ayurveda

The term *Ayurveda* comes from the Sanskrit words *ayus* (long life) and *veda* (knowledge). It is used to describe an ancient South Asian (primarily Indian) humoral medical system based around the inter-relationship between the body's 'humors' *vata* (wind), *pitta* (bile) and *kapha* (phlegm), its tissues and waste products.

In *Ayurveda*, food plays a critical role in maintaining the strength and stability of a body 'matrix' and foods with different but complementary qualities can be used to achieve humoral balance. The *nadis* or channels through which *prana* ('life force') is said to circulate connect via energy 'centers' called chakras.

A meditating figure illustrating the *chakras* – or energy centers and the *kundalini* - the source of 'primal energy' located in the spine. (Wellcome Library CC BY 4.0)

Similarly to TCM, *Ayurveda* has seen resurgence in use since the early 20th century and, through the combined efforts of pharmacologists such as R.N. Chopra, has been absorbed into 'mainstream' Indian medical practice.

Many regional or local variants of *Ayurveda* (e.g. *Siddha*) are still practiced, but *Ayurveda* has not been standardized or accepted to the same extent as TCM, despite efforts to promote and regulate it through AYUSH - the Indian government department of Ayurveda, Yoga and Naturopathy, Unani, Siddha and Homoeopathy.

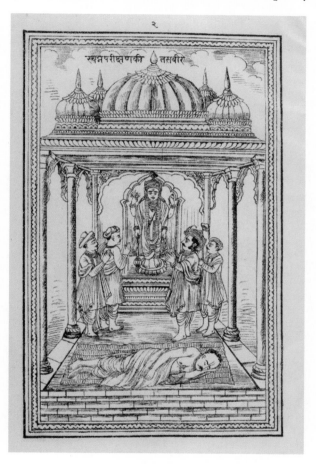

Worshippers praying to Dhanvantari, the God of Ayurvedic medicine for the health of a sick man in the foreground. (Wellcome Library CC BY 4.0)

Certain Ayurvedic scholars have suggested that *Ayurveda* is actually a philosophical framework rather than a 'system' and that, being predicated on harmony with nature; it is applicable to any country's medical practices.

Unani

The name *Unani* comes from the Arabic word for 'Greek' and is a system resulting from the assimilation of Greek medical philosophy into medical traditions from the Near and Middle East. *Unani*'s conceptual foundations lie in the Hippocratic notion of the four humors. From the 12th century onwards *Unani* was brought into India via the Arab conquests and it is widely practiced today across India, Bangladesh, Iran, Pakistan, and South Asia.

Abū ʿAlī al-Ḥusayn ibn ʿAbd Allāh ibn Sīnā (980-1037) known to the Western world as 'Avicenna' put forth the principles of *Unani* in his *Canon of Medicine* (1025).

'Avicenna' and 'pupil' (Wellcome Collection. CC BY 4.0)

In a way the *Canon of Medicine* became a vehicle for the humoral belief system in the West and Avicenna's work was widely consulted by European physicians up until the 17th century.

In *Unani*, or *Unani tibb* (medicine) as it is commonly known, health is believed to be regulated by the *akhlat* (humors); *arkan* (elements); *mizaj* (temperament); *a'za* (organs); *arwah* (vital spirit); *quwa* (powers) and *af'aal* (functions). As is the case with *Ayuvrveda* and TCM, *Unani* stresses the importance of physical exercise (*harakat wa sukoon al-baden*) and diet (*al-maakool wa al-mashroob*) to maintain good health.

Drawing of viscera etc. from Avicenna's, *Canon of Medicine.* (Wellcome Collection. CC BY 4.0)

A practitioner of *Unani* is called a *hakim* (Arabic for wise man or physician). A *hakim* diagnoses illnesses and formulates medicines for their patients. *Unani* has, in recent years, become more commercialized and elements of it have seeped into alternative and complementary medical practice.

Chapter Resources

Traditional Chinese Medicine Database @Taiwan

http://tcm.cmu.edu.tw

The Association of Traditional Chinese Medicine and Acupuncture (ATCM)

https://www.atcm.co.uk/

The Register of Chinese Herbal Medicine

http://rchm.co.uk/

Department of Ayurveda, Yoga and Naturopathy, Unani, Siddha and Homoeopathy (AYUSH)

http://ayush.gov.in/about-the-systems

Regulation

Geography: four types of globe. Engraving. (Wellcome Collection. CC BY 4.0)

Global Regulation

Europe

United Kingdom

United States

Global Regulation

Regulation of the natural medicines market exists at an international, national and regional level. The **World Health Organization** (WHO) - the specialized health agency of the United Nations – and pan-national bodies such as the **Pan-American Health Organization** (PAHO) deliver and promote WHO health strategy in their respective countries. Many countries also produce **pharmacopeia** which specify the standards required for the production of licensed medicines in their territories. A global overview of the regulatory/advisory framework is shown below.

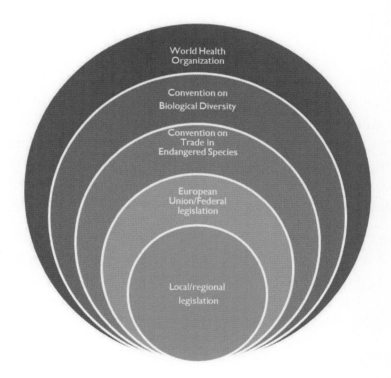

Some examples of measures which affect the global and national regulation and marketing of natural medicines are described overleaf.

Good Manufacturing Practices (GMP) adopted in 1968, the GMP program defines quality measures for all stages of production and quality control of 'medicinal products'.

The **Convention on Biological Diversity** (CBD) of 1992 promotes *"the conservation of biological diversity, the sustainable use of its components, and the fair and equitable sharing of benefits arising from the use of genetic resources"*. The **Nagoya Protocol** (2014) has enhanced aspects of the CBD regarding access and benefit-sharing.

Convention on International Trade in Endangered Species of Wild Fauna and Flora (CITES) of 1975 is an international agreement between governments to ensure that international trade in specimens of wild animals and plants does not threaten their survival.

Europe

The European Directive on Traditional Herbal Medicinal Products (THMPD) was implemented throughout the European Union (EU) in 2011 in order to provide a unified regulatory approval process for traditional herbal medicines. It has led to two kinds of marketing registrations: traditional use registrations (TURs) and well-established use marketing authorizations (WEU-MAs).

Under TUR, manufacturers must demonstrate that an herbal medicine has been used in the EU for at least 30 years, or 15 years in the European Economic Area (EEA) and 30 years outside it. WEU-MAs must demonstrate the same evidence of safety and effectiveness as any other pharmaceutical drug.

United Kingdom

Before 1997 the herbal medicines market in the UK was largely un-regulated; after a number of deaths involving herbal medicines, the Medicines and Healthcare products Regulatory Agency (MHRA) began to develop a partial regulatory system. This regulatory framework was bolstered by the introduction of the EU THMPD Directive which was transposed into UK legislation in 2005. To date, the MHRA has registered over 350 products under its traditional herbal registration (THR) scheme (see Appendix 1). The MHRA, in partnership with law enforcement agencies, has also carried out high profile seizures of fake or sub-standard medicines.

United States

In the US, a natural medicine can be classed as a biological product, cosmetic, drug, device, or dietary supplement. Dietary supplements cannot be marketed for the purpose of treating, diagnosing, preventing, or curing diseases, but they can be sold without FDA approval – however, the manufacturer is ultimately responsible for their safety and efficacy.

Drugs, as defined by the FDA are *"articles intended for use in the diagnosis, cure, mitigation, treatment, or prevention of disease"* and *"articles (other than food) intended to affect the structure or any function of the body of man or other animals"* and include prescription drugs (both brand-name and generic) and non-prescription (over-the-counter) drugs. Registered Naturopathic doctors may prescribe certain non-approved drugs although, in the case of certain conditions (e.g. cancers), it is illegal to prescribe non-approved drugs.

The Federal Food, Drug, and Cosmetic Act and Public Health Service Act also control dietary supplements and two licenses have been granted for a further FDA-proscribed category of **botanical drugs**, i.e. Sinecatechins - a water extract of green tea leaves from *Camellia sinensis* L., marketed as Veregen® for the treatment of genital warts, and, Mytesi™ – containing crofelemer from the latex of the *Croton lechleri* Müll.Arg. tree – used as an anti-diarrheal for HIV/AIDS patients undergoing anti-retroviral therapy (ART).

Chapter Resources

Global

Convention on Biological Diversity (CBD)

https://cbd.int/

Convention on Trade in Endangered Species

https://cites.org/

World Health Organization Good manufacturing practice (GMP)

https://who.int/medicines/areas/quality_safety/quality_assurance/gmp/en/

Europe

European Medicines Agency

https://ema.europa.eu

United Kingdom

UK Medicines and Healthcare products Regulatory Agency (MHRA)

https://gov.uk/government/organisations/medicines-and-healthcare-products-regulatory-agency

United States

US Food & Drug Administration

https://fda.gov/

Pan American Health Organization (PAHO)

https://www.paho.org/hq/

Case Studies

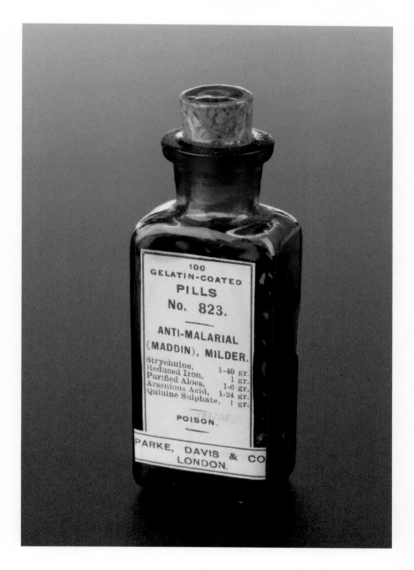

A bottle of anti-malarial pills containing quinine sulfate – derived fron the bark of *Cinchona* L. London, England, 1891-94. (Science Museum, London. CC BY 4.0)

Aristolochia

Aspirin

Cannabis

Galantamine

Paclitaxel

PHY906

St John's wort

Vinblastine and Vincristine

Aristolochia – Natural isn't Always Best

Aristolochia clematitis (H. Zell CC BY 3.0/GFDL)

aristolochic acid - $C_{17}H_{11}NO_7$

Overview

The Aristolochiaceae family contains around 500 species and, numerous *Aristolochia* species such as *Aristolochia fangchi* and *Aristolochia manshuriensis* were widely used medically before the discovery that they are highly toxic.

History

Aristolochia clematitis is known historically in European folk medicine as a birth aid. In TCM, a range of *Aristolochia* species are still used. In the 1990s a clinic in Belgium came under investigation after cases of renal (kidney) failure involving patients who had been prescribed *Stephania tetrandra (Han Fang Ji)* as a weight loss aid and had instead received *Aristolochia fangchi (Guang Fang Ji)*. Investigation established that *Aristolochia fangchi* contained highly nephrotoxic and oncogenic aristolochic acid. Aristolochic acid analogs (AAAs) are also found in the related *Asarum* genus.

Mode of Action

Aristolochic acid causes nephropathy (kidney disease) by kidney shrinkage and DNA mutations which cause damage to the proximal tubular cells and lead to ischemia (loss of blood supply) resulting from the thickening of endothelial walls in the kidney's arterioles (small-diameter blood vessels).

Current Status

All *Aristolochia* species are banned from sale in Europe. *Aristolochia Fangchi* has been banned in mainland China, Hong Kong and Taiwan and strict controls have been placed on the sale of other *Aristolochia* species in these markets.

Aspirin – An Ancient Rediscovery

Salix alba L. (willow) the source from which Aspirin is derived (CC BY-SA 2.5)

Aspirin (acetylsalicylic acid - $C_9H_8O_4$)

Overview

Aspirin is one the world's most widely used drugs with up to 120 billion pills consumed each year (Jones 2015). Aspirin is derived from the bark of the willow *Salix alba* L.

History

Evidence of willow bark as a remedy for fever dates back to ancient Sumer, Assyria and Egypt. In 18th century Europe willow bark became popular as a homeopathic remedy. Edward Stone, an

English vicar, carried out his own experiments on willow bark and discovered its active ingredient salicylic acid. Research by the German Pharmaceutical giant Bayer in 1897 led to the isolation and synthesis of acetylsalicylic acid – a modified version of salicylic acid with fewer side effects. It is marketed as the drug Aspirin to this day.

Method of Use

Prescription medication

Mode of Action

Aspirin is classed as an NSAID (non-steroidal anti-inflammatory drug) and works by suppressing the activity of the enzyme cyclooxygenase (COX) which, in turn, prevents prostaglandins (PGs) from causing inflammation, swelling, pain and fever.

Current Status

Legal, widely used, generic drug

Cannabis - A Popular Pariah

Cannabis sativa L. plant (CC 3.0 BY Chmee2)

cannabidiol (CBD)   δ9-trans-tetrahydrocannabinol (THC)

Overview

Cannabis (*Cannabis sativa* L. and *Cannabis indica* Lam.) are flowering plants from the Cannabaceae family. Cannabis species are often smoked or ingested and their stem fibers used for hemp cultivation.

History

Cannabis, known widely as *marijuana*, has been used recreationally and medicinally for centuries and hemp - a fiber from the stem of cannabis has a wide range of uses from textile to paper production.

Although illegal to grow, sell or consume in most countries, the last decade has seen an explosion in the use of 'medical *marijuana* and its controlled sale in parts of the United States.

Method of Use

Smoked, vaporized, cooked or oils

Mode of Action

Cannabinoids interact with cannabinoid, opioid and benzodiazepine receptors in the brain. Psychoactive tetra hydra cannabinols (THCs) and non-psychoactive cannabidiols (CBDs) act on pathways involving pain and inflammation and induce an analgesic effect.

Much research into Cannabis' therapeutic potential is now focused on the endocannabinoid system (ECS) – a network of retrograde neurotransmitters - which play a role in appetite, mood and memory regulation.

Current Status

UK: Class B drug, 1x licensed therapy (Sativex®) for MS patients. 1 (Epidiolex®) undergoing clinical trials. **US:** Schedule 1 prohibited drug. 1 x licensed therapy (Cesamet®)

Galantamine – A Reawakened Memory

Galanthus spp. (snowdrop) (David Paloch CC BY-SA 3.0)

galantamine $C_{17}H_{21}NO_3$

Overview

Galantamine is an alkaloid extracted from the bulbs and flowers of *Galanthus caucasicus* (Baker) Grossh. and *Galanthus woronowii* Losinsk. - from the Amaryllidaceae (onion) family.

History

Galanthus species (snowdrop) have a long record of folk medicinal use in parts of Europe as anti-inflammatories and memory enhancers. Maud Grieve, the renowned English herbalist, referred to historic accounts of Galanthus' *"digestive, resolutive and consolidante"* properties.

Mode of Action

In the brains of Alzeheimer's sufferers, the availability of acetylcholine - a crucial neurotransmitter - is reduced and healthy brain function impaired. Galantamine is metabolized by cytochrome P450 enzymes and appears to act by inhibiting the acetylcholinesterase (AChE) enzyme which breaks down acetylcholine. This leads to a greater availability of acetylcholine in the brain and some improvement in cognitive function for many patients.

Method of Use

Prescription medication, Galantamine hydrobromide capsule (FDA warning)

Current Status

US: approved by the Food and Drug Administration as safe and effective for the treatment of mild to moderate dementia. FDA warning for certain patients.

Paclitaxel – The Cancer Tree

Taxus baccata L. (Didier Descouens CC BY 4.0)

paclitaxel $C_{47}H_{51}NO_{14}$

Overview

Paclitaxel is a tetracyclic diterpenoid compound originally derived from the bark of the Western Pacific yew tree *Taxus brevifolia* Nutt. Paclitaxel - sold as Taxol® - is an important anti-cancer drug which destroys ovarian, breast and lung, bladder, prostate, melanoma, esophageal and solid tumor cancers.

History

Paclitaxel was identified for its anti-cancer properties in 1962 via the NCI screening program, in which it was shown to be toxic to living cells. From 1967-1993, almost all paclitaxel was derived from bark from the Pacific yew. As result of damage to Pacific yew populations, it is now produced by a semi-synthetic process from the needles and twigs of the closely related *Taxus baccata* L. or common yew.

Mode of Action

Paclitaxel works by tubulin inhibition. Tubulin is an integral part of a cell's structure and without it cancer cells cannot form a cytoskeleton and are therefore unable to divide and multiply.

Method of Use

Prescription medication, capsule.

Current Status

Legal, generic

PHY906 – A Mixed Blessing

L-R *Glycyrrhiza uralensis* Fisch; *Paeonia lactiflora* Pall (Ulf Eliasson CC BY-SA 3.0); *Scutellaria baicalensis* Georgi *(Stanislav Doronenko); Ziziphus jujuba* Mill. (CC BY 3.0)

Overview

Based on *Huangqin Tang*, a traditional Chinese herbal mixture for gastrointestinal health, dating back at least 1,800 years, PHY906 contains: *Glycyrrhiza uralensis* Fisch, *Paeonia lactiflora* Pall, *Scutellaria baicalensis* Georgi, and *Ziziphus jujuba* Mill. PHY906 has been developed as an adjuvant for patients undergoing chemotherapy. The synergy of numerous compounds in PHY906 has been shown to help reduce the side effects from chemotherapy.

History

PHY906 has been developed with the aid of PhytomicsQC - a novel method involving LC-MS, bio-response fingerprinting and *in vivo* pharmacology. A statistical method, the Phytomics Similarity Index (PSI), ensures quality control between batches of PHY906 by assessing the consistency of the chemical and bio-response markers, morphology and purity.

Method of Use Pharmaceutical drug

Mode of Action

PHY906 works to repair gut damage caused by conventional chemotherapy. In tests rodents undergoing chemotherapy which were also given PHY906 had fewer inflammatory immune cells (macrophages) than those treated with chemotherapy alone. Other inflammation markers — COX2, NF-κB and iNOS — were also reduced.

Current Status TBC

St John's wort – An Herbal Headache

Hypericum perforatum L. (Rowan McOnegal CC BY-NC)

L-R: hyperforin - $C_{35}H_{52}O_4$; hypericin – $C_{30}H_{16}O_8$

Overview

St John's wort generally refers to *Hypericum perforatum* L. - a flowering plant from the Hypericaceae family – now most often used as a mood enhancer or mild anti-depressant.

History

St John's wort has a long history of medicinal use and was used traditionally, in accordance with the doctrine of signatures, to treat skin conditions - based on the broken or perforated appearance of its leaves.

Mode of Action

Hypericin and hyperforin are naphthodianthrones which have been found to inhibit the uptake of serotonin – in a similar manner to Prozac® and other pharmaceutical selective serotonin reuptake inhibitors (SSRIs). St John's wort also increases dopamine and norepinephrine (adrenalin) levels in the brain. Used with SSRIs St John's wort can dangerously elevate serotonin levels, leading to serotonin syndrome – a potentially life-threatening condition.

St John's wort is an inducer of hepatic (liver) CYPs (cytochrome enzymes). CYPs (including cytochrome P450) metabolize many prescription medicines and, taken together with St John's wort, this may lead to slowing down in metabolism and a dangerous build-up of drugs in the body. Consequently, there have been multiple warnings about interactions with prescription drugs e.g. anti-cancer drugs, HIV drugs and anti-coagulants (blood thinners).

Current Status

UK: THR US: dietary (herbal) supplement

Vinblastine and Vincristine – *Veni, Vidi... Vinca*

Catharanthus roseus L. (Madagascar periwinkle) (Sue Snell CC BY 4.0)

L-R: vinblastine $C_{46}H_{58}N_4O_9$; vincristine $C_{46}H_{56}N_4O_{10}$

Overview

Vinblastine and Vincristine are widely used cancer chemotherapy drugs containing 'vinca' alkaloids isolated from *Catharanthus roseus* L., a flowering plant from the Apocynaceae family - commonly known as the Madagascar periwinkle.

History

The periwinkle has a long record of use as an astringent, tonic, cure for scurvy and treatment for diabetes. In the 1950s researchers investigating its apparent anti-diabetic properties discovered that animals dosed with *Catharanthus roseus* L. would show a reduced white blood cell count after treatment.

Following these experiments *Catharanthus roseus* L., was added to the U.S. National Cancer Institute's screening program and was developed into two successful cancer chemotherapy drugs – comprising analogs of sulfated salts of vinca alkaloids and prescribed for conditions such as childhood leukemia and non-Hodgkin's Lymphoma.

Mode of Action

Both drugs inhibit microtubule formation, by binding to tubulin, therefore preventing cell mitosis - but other mechanisms may also exist.

Current Status

Legal, prescribed cancer chemotherapy drugs

Chapter Resources

Drugs

https://fda.gov/Drugs/

Poisonous Plants

FDA Poisonous Plant Database

https://accessdata.fda.gov/scripts/plantox/index.cfm

Conclusion

Traffic at a crossroads. (Yu Sung Lak CC BY-SA 2.0)

Conclusion

Globally, natural medicine has reached a crossroads - the growing availability of natural medicines online and the worrying trend of fake medicines flooding into the market, have all contributed to increased concern about safety and efficacy.

A recent study by the UCL School of Pharmacy, on Ginkgo, Milk Thistle and *Rhodiola rosea* products, discovered that 20 - 40% of products tested were mislabelled or contained adulterants and this may be just the 'tip of the iceberg'. Even analyses of batches of apparently standardized commercial products (e.g. St John's wort) show markedly different chemical profiles.

Drug approvals and market share for natural or natural-derived medicines are increasing and, as a result, governments and health regulators are struggling to keep up with the unique challenges posed by natural medicines.

Yet, there are positive developments too - on the world stage, natural medicines now have their foot firmly in the door. In 2015, Professor Youyou Tu, Chief Scientist at the China Academy of Traditional Chinese Medicine, was awarded the Nobel Prize for Medicine, for her pioneering work on artemisinin - a compound isolated from sweet wormwood (*Artemesia annua* L.) – a drug which is of immense value in the global fight against malaria.

Initiatives such as the **MHRA yellow card scheme** and the **#FakeMeds campaign** in the UK have put safety checks on natural medicines on a footing with prescribed over the counter medicines, by encouraging consumers to report side effects and quality concerns. In the United States, the FDA operate a similar scheme called the **MedWatch Adverse Event Reporting program**.

There is also a renewed interest in the therapeutic potential of 'drugs of abuse' – encompassing an array of natural products which have become better known for their 'recreational' rather than purely medical use. Whereas these drugs were once labelled illegal and dangerous, serious research is now focused on finding their possible value as medicines. For instance psilocybin - a compound derived from 'magic' mushrooms from the genus *Psilocybe* - has shown promise as an anxiolytic for patients with end-stage cancer.

Both *Ayahuasca*, a South American shamanic medicinal mixture and ibogaine, an alkaloid isolated from the root bark of the *Tabernanthe iboga* - used ritually in traditional African medicine - have been investigated for their ability to combat drug addiction.

Then there is Cannabis - perhaps the most misunderstood of natural medicines. In the last decade the legalisation movement has grown stronger, as has the popularity of hemp seed and CBD oils. Sativex®, an oral spray for patients with multiple sclerosis, has received marketing authorization in many countries and, its producer also recently announced successful phase III trials of Epidiolex® - a cannabidiol (CBD) drug for the prevention of epileptic seizures in a rare and severe form of childhood-onset epilepsy. In the US nabilone, a synthetic cannabinoid - is now marketed as the anti-emetic drug Cesamet®.

However, new legislative moves in the UK – such as the **Psychoactive Substances Act** 2016 have called into question the legality of any psychoactive substance - tea and coffee included... In addition, the withdrawal of public funding for all homeopathic medicines and, many herbal medicines, from 2018 - will change the landscape of natural medicines research and practice.

Still, In spite of the complexities, political or otherwise, of producing new natural medicines or reviving traditional remedies, these are exciting times for the world of natural medicine and, I believe, in this 'information era', that we owe it to ourselves to learn more about the risks and benefits natural medicines present. The desire to use natural medicines may stem from a sense that conventional medicine is impersonal or not patient-centered; it may reflect a desire for medicines which treat the patient rather than the condition. Certainly, the medicines we take are detached from their origins; when we look at a modern pharmaceutical product it could equally be a placebo pill. This distancing from our environment is hugely at odds with the way humans have found healing remedies throughout history.

Although CAM has become a flag bearer for all things natural, traditional medicines must be understood within their historic context where they are not simply 'things' to prevent or cure illness but, to borrow modern jargon, 'to be taken as part of a healthy life style'. Perhaps, by understanding the lifestyles espoused by traditional medical systems, rather than simpy viewing their remedies in isolation, we can re-connect with this insight.

There is much that has been said historically that modern science validates. The old idiom *"One man's meat is another man's poison"*, attributed to Lucretius the Roman poet and philosopher, could be taken as an early call for personalized medicine, a concept that the mainstream medical community is just beginning to embrace. Paracelsus' famous words *"All things are poison and nothing is without poison; only the dose makes a thing not a poison."* could be construed similarly. No matter how wide the natural *versus* pharmaceutical medicines divide might be, the salient principles of traditional knowledge seem inarguable, namely, that:

- **Health and lifestyle are inseparable**

- **We must resepect nature**

- **The best medicine for a condition must take into account the patient's circumstances - physical, emotional, economic and geographic**

...If further reason is needed to take note of these principles, It is worth contemplating that, without the medical traditions our ancestors devised, human beings might not be here today.

L: A South African medicine man or *shaman*. Colour process print after a photograph by G.W. Wilson. (Wellcome Collection. CC BY 4.0) R: A *Curandera* (healer) performing a *limpieza* (cleansing ritual) in Cuenca, Ecuador. (calliopejen CC BY-SA 2.0)

Of course, we must challenge the supposition that a medicine is safe and effective just because it is natural, but equally its opposite that a medicine is unsafe and ineffective just because it is natural. It seems hard to dispute that the best medicine for a patient is the one that will provide the best health outcome.

Natural medicines may not be intrinsically better or worse but - with the appropriate controls and safety measures - they can be just as safe and effective. As I said at the beginning, all medicines have their potential risks and benefits. I would like to see a situation where people are clearer about the risks and benefits of *any* medicine that they are taking – natural or otherwise – and are not afraid to seek out the facts. I hope that this book goes some way towards encouraging this.

Appendix 1.

MHRA Guidance on Herbal medicines granted a traditional herbal registration

MHRA Guidance on Herbal medicines granted a traditional herbal registration (THR) in the United Kingdom – February 2018*

*Please note that this list has been modified to exlude instances where multiple licenses have been granted for the same active ingredient(s) and product indications/traditional use.

Active ingredient	Product indication - Traditional use
Agnus castus (Vitex agnus-castus L.)	For the relief of premenstrual symptoms such as irritability, mood swings, breast tenderness, bloating and menstrual cramps, based on traditional use only.
Agnus Castus (Vitex agnus-castus L.) fruit	To help relieve symptoms associated with premenstrual syndrome, such as irritability, mood swings, breast tenderness, bloating and menstrual cramps. This is based on traditional use only.
Agnus Castus Fruit (Vitex agnus-castus L.)	For relief of the symptoms associated with premenstrual syndrome
Agnus Castus Fruit (Vitex agnus-castus L.)	For the relief of the symptoms associated with premenstrual syndrome
Agnus Castus Fruit (Vitex agnus-castus L.)	To help relieve premenstrual symptoms such as irritability, mood swings, breast tenderness, bloating and menstrual cramps, based on traditional use only. Agnus castus fruit (Vitex agnus castus L., fructus)
Agnus Castus Fruit (Vitex agnus-castus L.)	To help relieve premenstrual symptoms such as irritability, mood swings, breast tenderness, bloating and menstrual cramps, based on traditional use only.

Angelica root (Angelica archangelica L.), Cinnamon bark (Cinnamon zeylanicum Nees.), Nutmeg seed (Myristica fragrans Houtt.) Melissa leaf (Melissa officinalis L.)	for the symptomatic relief of nausea, stomach ache and upset stomach, based on traditional use only.
Aniseed (Pimpinella anisum L.) Elderflower (Sambucus nigra L.) Iceland Moss thallus (Cetraria islandica L.), Marshmallow root (Althaea officinalis L.), Thyme herb (Thymus vulgaris L & Thymus zygis L), White Horehound herb (Marrubium vulgare L)	For the symptomatic relief of dry and irritating coughs, based on traditional use only.
Arnica (Arnica montana L). Birch (Betula pendula Roth) leaf	For the symptomatic relief of rheumatic pain, muscular pain and stiffness, backache, fibrositis, bruising, cramp, sprains and minor sports injuries, based on traditional use only.
Arnica (Arnica montana L.) fresh, whole plant	For the relief of minor sports injuries, bruises, muscular pain, stiffness and sprains, based on traditional use only.
Arnica Flower (Arnica montana L.)	For the symptomatic relief of muscular aches and pains, stiffness, sprains, bruises and swelling after contusions.
Arnica flowers (Arnica Montana L.)	For the symptomatic relief of bruises, based on traditional use only.
Arnica flowers (Arnica Montana L.)	For the symptomatic relief of bruises, based on traditional use only.
Arnica (Arnica montana L.) fresh, whole plant	For the relief of muscular aches, pain, stiffness, sprains, bruises, swelling after contusions and minor sports injuries based on traditional use only.

Artichoke (Cynara scolymus L.)	For the relief of digestive complaints, such as indigestion, upset stomach, nausea, feelings of fullness and flatulence (wind), particularly caused by over indulgence of food and drink, based on traditional use only.
Artichoke (Cynara scolymus L.) leaf	For the relief of digestive complaints, such as indigestion, upset stomach, nausea, feelings of fullness and flatulence (wind), particularly caused by over indulgence of food and drink, based on traditional use only.
Artichoke (Cynara scolymus L.) leaf	For the symptomatic relief of digestive disorders such as dyspepsia and flatulence based on traditional use only.
Artichoke leaf (Cynara scolymus L.)	For the relief of digestive complaints, such as indigestion, upset stomach, nausea, feelings of fullness and flatulence, particularly caused by over indulgence of food and drink, based on traditional use only.
Artichoke leaves (Cynara scolymus L.), Dandelion root and herb (Taraxacum officinalis WEB,) Boldo leaves (Peumus boldus MOLINA.) Peppermint herb Mentha x piperita L.).	For indigestion, sensation of fullness and flatulence associated with over-indulgence in food or drink, or both, based on traditional use only.
Artichoke leaves (Cynara scolymus L., folium) Milk Thistle fruit (Silybum marianum (L.) Gaertn., fructus) Dandelion root and herb (Taraxacum officinalis WEB., radix cum herba) Boldo leaves (Peumus boldus MOLINA., folium)	A traditional herbal medicinal product used for indigestion, sensation of fullness and flatulence associated with over-indulgence of food or drink, or both, based on traditional use only.
Asparagus root (Asparagus officinalis L.)	To increase the amount of urine for the purpose of flushing the urinary tract to assist in minor urinary complaints, based on traditional use only.

Asparagus Root (Asparagus officinalis L.), Parsley Herb (Petroselinum crispum (MilL.) A. W. Hill)	To increase the amount of urine for the purpose of flushing the urinary tract to assist in minor urinary complaints, based on traditional use only.
Black Cohosh dry extract (cimicifuga racemosa (l) nutt.)	for the relief of symptoms of the menopause, such as hot flushes, night sweats, and temporary changes in mood. This is based on traditional use only. As there is evidence that black cohosh may have hormone-like actions, it should only be used by women of childbearing potential if contraception is used.
Black Cohosh rhizome and root (Cimicifuga racemosa (L) Nutt.)	For symptomatic relief of backache, muscular and rheumatic aches and pains. This is based on traditional use only.
Black Cohosh rhizome and root (Cimicifuga racemosa (L.) Nutt.)	For the relief of symptoms of the menopause, such as hot flushes, night sweats, and temporary changes in mood (such as nervous irritability and restlessness) based on traditional use only.
Black Cohosh rhizome and root (Cimicifuga racemosa (L.) Nutt.).	For the relief of symptoms of the menopause, such as hot flushes, night sweats, and temporary changes in mood (such as nervous irritability and restlessness)
Black Cohosh rhizome and root (Cimicifuga racemosa (L.) Nutt.).	For the relief of Backache, rheumatic aches and pains, and pain in the muscles and joints, based on traditional use only.
Bladderwrack (Fucus vesiculosus L.)	As an aid to slimming as part of a calorie controlled diet, based on traditional use only

Bladderwrack (Fucus vesiculosus L.), Clivers (Galium aparine L.)	As an aid to slimming as part of a calorie controlled diet, based on traditional use only.
Bladderwrack thallus (Fucus vesiculosus L. or Fucus serratus L)	As an aid to slimming as part of a calorie controlled diet, based on traditional use only
Bladderwrack Thallus (Fucus vesiculosus L.)	As an aid to slimming as part of a calorie controlled diet, based on traditional use only.
Bladderwrack thallus (Fucus vesiculosus L.) Dandelion root (Taraxacum officinale Weber ex Wigg) Boldo leaf (Peumus boldus Molina) Butternut bark (Juglans cineraria L.)	As an aid to slimming as part of a calorie controlled diet, based on traditional use only.
Bladderwrack thallus (Fucus vesiculosusL.) Burdock root (Arctium lappa L.) Ground Ivy leaf (Glechoma hederaceaL.)	To relieve symptoms in mild cases of water retention, based on traditional use only.
Boldo leaf (Peumus boldus L.) Dandelion root (Taraxacum officinale Weber ex Wigg.) Uva Ursi leaf (Arctostaphylos uva ursi L.) Bladderwrack thallus (Fucus vesiculosus L.)	To relieve symptoms in mild cases of water retention, based on traditional use only.
Boldo leaf (Peumus boldus Molina), Celery seed (Apium graveolens L.)	To relieve bloating associated with premenstrual water retention, based on traditional use only.
Boneset (Eupatorium perfoliatum L.) herb. Hyssop (Hyssopus officinalis L.) herb. Burdock (Arctium lappa L.) root	to relieve the symptoms of nasal catarrh and catarrh of the throat, based on traditional use only.
Buchu leaf (Agathosma betulina (Berg), Pillans Uva Ursi leaf (Arctostaphylos uva-ursi (L.) Spreng. Dandelion root (Taraxacum officinale Weber ex Wigg.)	To relieve bloating associated with premenstrual water retention, based on traditional use only.
Burdock (Arctium lappa L.) root	To increase the amount of urine for the purpose of flushing the urinary tract to assist in minor urinary complaints, based on traditional use only. For the relief of seborrhoeic skin disorders, such as flaky skin or dandruff. This is based on traditional use only.

Burdock (Arctium lappa L.) root, Sarsaparilla (Smilax aristolochiaefolia) root, Red Clover (Trifolium pratense) flower, Queen's Delight (Stillingia sylvatica L.) root, Cascara (Rhamnus purshiana L.) bark, Poke (Phytolacca decandra) root, Prickly Ash (Zanthoxylum americanum) bark	For the symptomatic relief of minor skin conditions causing skin irritation, including allergic conditions, dermatitis, mild acne and mild eczema based on traditional use.
Burdock Root (Arctium lappa L)	To increase the amount of urine for the purpose of flushing the urinary tract to assist in minor urinary complaints. This is based on traditional use only
Burdock root (Arctium lappa L), Hyssop herb (Hyssopus officinalis L)	To relieve the symptoms of nasal catarrh and catarrh of the throat based on traditional use only.
Burdock root (Arctium lappa L.) Blue Flag root (Iris versicolor L.) Queen's Delight root (Stillingia sylvatica L.) Sarsaparilla root (Smilax spp.)	For the symptomatic relief of mild eczema, based on traditional use only.
Burdock root (Arctium lappa L.) Fumitory herb (Fumaria officinalis L.)	For the symptomatic relief of minor skin conditions such as spots, pimples and blemishes, based on traditional use only.
Burdock root (Arctium lappa L.), Blue Flag Iris rhizome (Iris versicolor L.), Sarsaparilla root (Smilax species: S. regelii Killip et Morton; S. febrifuga Kunth)	For the symptomatic relief of minor skin conditions such as spots, pimples, blemishes, mild acne, and mild eczema, based on traditional use only.
Butcher's Broom rhizome (Ruscus aculeatus L., rhizome)	To relieve symptoms of discomfort and heaviness of the legs related to minor venous circulatory disturbances, based on traditional use only. relieve symptoms of itching and burning associated with haemorrhoids, based on traditional use only.
Butternut bark (Juglans cineraria L.) Dandelion Root (Taraxacum officinale Weber ex Wigg.). Boldo leaf (Peumus boldus Molina) Bladderwrack thallus (Fucus vesiculosus L.)	As an aid to slimming as part of a calorie controlled diet, based on traditional use only.

Calendula flowers (Calendula officinalis L.)	For the symptomatic relief of sore and rough skin, based on traditional use only.
Calendula flowers (Calendula officinalis L.)	For the symptomatic treatment of minor inflammations of the skin and as an aid in healing of minor wounds, based on traditional use only.
Calendula officinalis herb	For minor wounds, cuts and grazes, based on traditional use only.
Calendula officinalis herb (Calendula officinalis L. Herba).	For the relief of minor wounds, cuts and grazes, based on traditional use only.
Capsicum Cajuput oil Camphor Turpentine	For the symptomatic relief of rheumatic or muscular pain, sprains, stiffness and lower backache including sciatica and lumbago, based on traditional use only
Capsicum fruit (Capsicum frutescens L.) Guaiacum resin (Guaiacum officinale L) Blue Flag rhizome (Iris versicolor L.) Uva ursi leaf (Arctostaphylos uva-ursi L.)	For the relief of backache, rheumatic pain and general aches and pains in the muscles and joints, based on traditional use only.
Celery (Apium graveolens L.) seed	For the relief of rheumatic aches and pains, based on traditional use only.
Celery Seed (Apium graveolens L.)	For the relief of rheumatic aches and pains, based on traditional use only.
Centaury Herb (Centaurium erythraea Rafn.), Rosemary Leaf (Rosmarinus officinalis L.), Lovage Root (Levisticum officinale Koch.)	To help flushing of the urinary tract and to assist in minor urinary complaints associated with cystitis in women only, based on traditional use only (Century, Rosemary & Loveage)
Clove oil (Syzygium aromaticum (L.) Merill et L. M. Perry)	For the temporary relief of toothache, based on traditional use only.

Cola seed (Cola nitida (Vent)., Schott et EndL. seed)	For symptoms of temporary fatigue, weakness and exhaustion based on traditional use only.
Cola seed (Cola nitida (Vent. (Schott et EndL.) Damiana leaf (Turnera diffusa WilL.) Saw Palmetto fruit (Serenoa repens (Bartram) Small)	To help relieve symptoms of fatigue and weakness , based on traditional use only.
Cola seed (Cola nitida (Vent.) Schott et EndL.) Damiana leaf (Turnera diffusa WilL.) Saw Palmetto fruit (Serenoa repens (Bartram))	To relieve fatigue, exhaustion and debility, based on traditional use only.
Comfrey (Symphytum officinale L.) root	For the symptomatic relief of minor sprains and bruises, based on traditional use only.
Common Rue herb (Ruta graveolens L)	For the relief of muscular aches and sprains, based on traditional use only.
Corn Silk Herb (Zea mays L.) Damiana Leaf (Turnera diffusa WilL.) Saw Palmento Fruit (Serenoa repens (Bartram)	To relieve fatigue, exhaustion and debility, based on traditional use only.

Costus root (*Saussurea costus. (Falc.) Lipsch., radix*) Iceland moss (*Cetraria islandica. (L.) Ach. s.L., thallus*) neem fruit (*Azadirachta indica A. Juss., fructus*) cardamom fruit (*Elettaria. cardamomum (Roxb.) Maton var. minuscula Burk., fructus*) myrobalan fruit (*Terminalia chebula. Retz, fructus*) red sanderswood (*Pterocarpus. santalinus L.f., lignum*) allspice (*Pimenta dioica (L.) Merr., fructus*) Bengal quince (*Aegle marmelos. (L.) Corrêa, fructus*) European columbine (*Aquilegia. vulgaris L., herba*) liquorice root (*Glycyrrhiza glabra. L., radix*) ribwort plantain (*Plantago. lanceolata L. s.L., folium*) knotgrass (*Polygonum aviculare. L. s.L., herba*) golden cinquefoil (*Potentilla aurea. L., herba*) clove (*Syzygium aromaticum (L.). Merr. et L. M. Perry, flos*) kaempferia galanga rhizome (*Kaempferia galanga L., rhizoma*) heart leaf sida (*Sida cordifolia L., herba*) valerian root (*Valeriana officinalis. L. s.L., radix*) lettuce leaf (*Lactuca sativa var. capitata L., folium*) calendula flower head (*Calendula officinalis L., flos cum calyce*) solid fraction of crude camphor oiL. (*Cinnamomum camphora (L.) Sieb., aetherolei fractio solida*).	A traditional herbal medicinal product used to relieve the symptoms of Raynaud's syndrome and for the relief of symptoms associated with minor venous circulatory disturbances such as tired heavy legs, pain, swelling, and for calf cramps, based on traditional use only.
Couch Grass Herb (*Agropyron repens (L) Beauv.*) Marshmallow Root (*Althaea officinalis L.*) Buchu leaf (*Agathosma betulina (Berg) Pillans*) Uva Ursi leaf (*Arctostaphylos uva-ursi (L.) Spreng.*) Juniper berries (*Juniperus communis*) Clivers herb (*Galium aparine*)	To help flushing of the urinary tract and to assist in minor urinary complaints associated with cystitis in women only, based on traditional use only.
Cranberry fruit	For the relief of symptoms of minor lower urinary complaints associated with cystitis in women
Damiana leaf (*Turnera diffusa*) Cola seeds (*Cola nitida (Vent.) Schott et EndL.*) Lucerne herb (*Medicago sativa*) Saw palmetto fruit (*Serenoa repens (Bartram*) Uva ursi leaf (*Arctostaphylos uva-ursi*)	Fo help relieve symptoms of fatigue and weakness, particularly in older men, based on traditional use only.

Dandelion (Taraxacum officinale Weber) root	To relieve the symptoms of mild digestive disorders, such as dyspepsia and flatulence and temporary loss of appetite. To increase the amount of urine for the purpose of flushing the urinary tract to assist in minor urinary complaints. This is based on traditional use only.
Dandelion root	To relieve the symptoms of mild digestive disorders, such as dyspepsia and flatulence and temporary loss of appetite. Increase the amount of urine for the purpose of flushing the urinary tract to assist in minor urinary complaints. This is based on traditional use only.
Dandelion root (Taraxacum officinale), Uva-ursi leaf (Arctostaphylos uva-ursi L.), Parsley piert herb (Alchemilla arvensis L.), Buchu leaf (agathosma betulina L.)	To relieve symptoms of mild water retention, based on traditional use only.
Devil's Claw (Harpagophytum procumbens (Burch.) DC ex Meissn)	For the relief of backache rheumatic or muscular pain and general aches and pains in the muscles and joints, based on traditional use only.
Devil's Claw (Harpagophytum procumbens (Burch.) DC. Ex Meissn)	For the relief of backache rheumatic or muscular pain and general aches and pains in the muscles and joints, based on traditional use only.
Devil's Claw (Harpagophytum procumbens (Burch.) DC. Ex Meissn)	For the relief of backache, rheumatic or muscular pain and general aches and pains in the muscles and joints. This is based on traditional use only.

Devil's Claw (Harpagophytum procumbens (Burch.) DC.ex Meissn))	For the relief of backache, rheumatic or muscular pains and general aches and pains in the muscles and joints
Devil's claw quantified dry extract (harpagophytum procumbens dc. And / or h.zeyheri L. Decne.)	For relief of backache, rheumatic or muscular pain, and general aches and pains in the muscles and joints, based on traditional use only.
Devil's Claw Root (Harpagophytum procumbens (Burch.) DC.ex Meissn)	For the relief of backache, rheumatic or muscular pains and general aches and pains in the muscles and joints
Devil's Claw Root (Harpagophytum procumbens (Burch.) DC.ex Meissn)	For the relief of backache, rheumatic or muscular pain, and general aches and pains in the muscles and joints, based on traditional use only.
Devil's Claw Root (Harpagophytum procumbens (Burch.) DC.ex Meissn)	Symptomatic relief of backache and muscular and rheumatic aches and pains, based on traditional use only.
Devil's Claw Root (Harpagophytum procumbens (Burch.) DC.ex Meissn)	For the relief of backache rheumatic or muscular pain and general aches and pains in the muscles and joints, based on traditional use only.
Devil's Claw root (Harpagophytum procumbens D.C. and /or H. zeyheri L. Decne, radix)	For the relief of backache, rheumatic or muscular pain, and general aches and pains in the muscles and joints based on traditional use only.
Echinacea angustifolia root	For the symptomatic relief of minor skin conditions such as spots, pimples and blemishes, based on traditional use only.
Echinacea angustifolia root, Wild Indigo root, Fumitory herb	To relieve: the symptoms of the common cold and flu type infections and minor skin conditions based on traditional use only.

Echinacea pallida (Nutt.) Nutt. Root Echinacea purpurea (L.) Moench.root	To relieve the symptoms of the common cold and influenza type infections, based on traditional use only.
Echinacea pupurea (L.)	To relieve the symptoms of the common cold and influenza type infections based on traditional use only.
Echinacea purpurea (echinacea purpurea (L.) Moench herb), Uva-ursi herb (arctostaphylos uva-ursi (L.) Spreng, Herb)	For the relief of minor urinary complaints associated with cystitis in women, such as burning sensation during urination and/or frequent urination, based on traditional use only.
Echinacea purpurea (L.) Moench	To relieve the symptoms of the common cold and influenza type infections. This is based on traditional use only.
Echinacea purpurea (L.) Moench	To relieve the symptoms of the common cold and influenza type infections based on traditional use only.
Echinacea purpurea (L.) Moench herb Echinacea purpurea (L.) Moench root Sage leaves (Salvia officinalis L. folium)	To relieve sore throats associated with coughs, colds and flu, based on traditional use only.
Echinacea purpurea (L.) Moench root	To relieve the symptoms of the common cold and influenza type infections. This is based on traditional use only.
Echinacea purpurea (L.) Moench root	To relieve the symptoms of the common cold, and influenza type infections based on traditional use only.
Echinacea purpurea fresh herb	To relieve the symptoms of the common cold and influenza type infections, based on traditional use only.

Echinacea purpurea Herb & Root (Echinacea purpurea (L.)	To relieve the symptoms of the common cold and influenza type infections based on traditional use only.
Echinacea purpurea Herb & Root (Echinacea purpurea (L.) Moench)	To relieve the symptoms of the common cold and influenza type infections based on traditional use only.
Echinacea purpurea Herb (Echinacea purpurea (L.) Moench)	To relieve the symptoms of the common cold and influenza type infections.
Echinacea purpurea radix tincture 1:1, Echinacea purpurea (L.)	For the relief of the symptoms of the common cold and influenza type infections based on traditional use only.
Echinacea purpurea Root (Echinacea purpurea (L.) Moench)	To relieve the symptoms of the common cold and influenza type infections based on traditional use only.
Echinacea purpurea Root (Echinacea purpurea (L.) Moench)	For the symptomatic relief of minor skin conditions, such as spots, pimples and blemishes based on traditional use only. (Echinacea purpurea root)
Echinacea purpurea Root (Echinacea purpurea (L.) Moench)	To relieve the symptoms of common cold and influenza type infections based on traditional use only.
Echinacea purpurea Root (Echinacea purpurea (L.) Moench)	To relieve the symptoms of the common cold and influenza type infections. This is based on traditional use only
Echinacea purpurea Root (Echinacea purpurea (L.) Moench)	To treat the symptoms of the common cold, influenza and minor upper respiratory tract infections based on traditional use only.

Echinacea purpurea Root (Echinacea purpurea (L.) Moench)	To relieve the symptoms of the common cold and influenza type infections based on traditional use only
Echinacea purpurea Root (Echinacea purpurea (L.) With Vitamin C and Zinc	To relieve the symptoms of the common cold and influenza type infections based on traditional use only. Echinacea with Vitamin C plus Zinc
Echinacea purpurea root (Echinacea, purpurea (L.) Moench).	To relieve the symptoms of the common cold and influenza-type infections based on traditional use only
Echinacea purpurea root dry extract (echinacea purpurea root)	For the relief of the symptoms of the common cold and influenza type infections, based on traditional use only.
Echinacea purpurea root(Echinacea purpurea (L.)	For the symptomatic relief of minor skin conditions such as spots, pimples, and blemishes, based in traditional use only.
Elder Flower (Sambucus nigra L.) Hemlock Spruce (Pinus canadensis L.) Bayberry Bark (Myrica cerifera L.)	To relieve of the symptoms of colds and flu, chills and sore throats, based on traditional use only.
Elder flowers (Sambucus nigra. L.), Bayberry bark (Myrica cerifera L.), Hemlock Spruce bark (Tsuga canadensis Carr.)	To relieve the symptoms of cold and flu, sore throats and chills, based on traditional use only.
Elm (Ulmus rubra MuhL.) bark powder	To relieve the symptoms of indigestion, heartburn and flatulence, based on traditional use only.
Eyebright quantified dry extract (euphrasia officinalis L.)	For the relief of blocked sinuses and catarrh. This is based on traditional use only.

Feverfew (tanacetum parthenium (L.) Schulz bip.)	For the prevention of migraine headaches based on traditional use only.
Feverfew Herb (Tanacetum parthenium (L.) Schultz Bip.)	For the prevention of migraine headaches
Fresh oat herb (Avena sativa L., herba rec.)	For the temporary relief of symptoms associated with mild stress such as mild anxiety and to aid sleep, based on traditional use only.
Garlic bulb (Allium sativum L.) bulb.	To relieve the symptoms of upper respiratory tract infections including common cold, cough, catarrh, blocked or runny nose, and sinus congestion based on traditional use only.
Garlic Powder (Allium sativum L.) bulb, Echinacea angustifolia root	To relieve the symptoms of catarrh, rhinitis and nasal congestion, based on traditional use only.
Gentian Root (Gentiana lutea L.), Verbena Herb (Verbena officinalis L.), Sorrel Herb (Rumex acetosa L.), Elderflower (Sambucus nigra L.), Primrose Flower with Calyx (Primula veris L.)	To relieve nasal congestion and sinusitis based on traditional use only. (Gentian root, Verbena herb, Garden Sorrel, Elderflower, Primrose)
Gentian Root (Gentiana lutea L.), Verbena Herb (Verbena officinalis L.), Sorrel Herb (Rumex acetosa L.), Elderflower (Sambucus nigra L.), Primrose Flower with Calyx (Primula veris L.)	To relieve nasal congestion and sinusitis based on traditional use only. Gentian root, Verbena herb, Garden Sorrel, Elderflower, Primrose)
German Chamomile flower (Matricaria recutita L. flower), Valerian root (Valeriana officinalis L. root), Scullcap herb (Scutellaria lateriflora L. herb).	For the temporary relief of symptoms associated with stress such as fatigue, exhaustion and mild anxiety based on traditional use only.

Ginger (Zingiber officinale ROSCOE) rhizome	To relieve the symptoms of minor digestive complaints such as indigestion, dyspepsia, feeling of fullness, flatulence and temporary loss of appetite. To relieve symptoms of travel sickness based on traditional use only.
Ginger (Zingiber officinale ROSCOE) rhizome	For symptomatic relief of travel sickness based on traditional use only.
Ginger (Zingiber officinale ROSCOE) rhizome	For symptomatic relief of minor digestive complaints such as indigestion, dyspepsia, feeling of fullness, flatulence and temporary loss of appetite based on traditional use only.
Ginger rhizome (Zingiber officinale Roscoe) Milk Thistle fruits (Silybum marianum L. Gaertner) Holy Thistle herb (Cnicus) Valerian root (Valeriana officinalis L.) Dandelion root (Taraxacum officinale Weber ex Wigg.) Myrrh resin (Commiphora myrrha EngL.)	For the relief of heartburn, stomach ache, indigestion, flatulence and dyspepsia, based on traditional use only.
Ginger rhizome (Zingiber officinale Roscoe), Capsicum fruit (Capsicum frutescens L.), Gentian root (Gentiana lutea L.)	For the symptomatic relief of indigestion, based on traditional use only.
Ginkgo biloba L. leaf	To relieve the symptoms of Raynaud's syndrome and tinnitus, based on traditional use only.
Ginkgo Leaf (Ginkgo biloba L.)	To relieve the symptoms of Raynaud's syndrome and tinnitus, based on traditional use only.
Ginseng Root (Panax ginseng C. A. Meyer)	For the temporary relief of fatigue, weakness and exhaustion, based on traditional use only.

Guaiacum resin (Guaiacum officinale L), Prickly Ash Northen bark (Zanthoxylum americanum Mill), Celery seed oil (Apium graveolens L oil)	For the relief of backache, rheumatic or muscular pain, and general aches and pains in the muscles and joints, based on traditional use only.
Guaiacum Resin (Guaiacum officinaleL.) Dandelion root (Taraxacum officinale Weber ex Wigg) Celery fruit (Apium graveolens L.) Buckbean herb (Menyanthes trifoliataL.)	For the relief of backache, rheumatic pain and general aches and pains in the muscles and joints, based on traditional use only.
Hamamelis Water (Hamamelis virginiana L.) Zinc Oxide Cade Oil (Juniperus oxycerus L.)	For the relief of minor skin inflammation and irritation due to varicose veins in the lower legs, based on traditional use only.
Hop strobile (Humulus lupulus L.) Scullcap herb (Scutellaria lateriflora L.) Vervain herb (Verbena officinalis L.) Valerian root (Valeriana officinalis L.)	A traditional herbal medicinal product used for the short term relief of symptoms associated with stress such as mild anxiety, tenseness or irritability, based on traditional use only.
Hop strobile (Humulus lupulus L.), Valerian root (Valeriana officinalis L.), Passion Flower herb (Passiflora incarnata L.)	for the temporary relief of sleep disturbances based on traditional use only.
Hop strobile (Humulus lupulus L.), Valerian root (Valeriana officinalis L.), Passion Flower herb (Passiflora incarnata L.)	Used for temporary relief of symptoms associated with stress, such as mild anxiety based on traditional use only.
Hop Strobiles (Humulus lupulus L.), Valerian root (Valeriana officinalis L.), Gentian root (Gentiana lutea L.)	For the temporary relief of; symptoms associated with stress such as mild anxiety and irritability including symptoms associated with menopause such as flushings and cold sweats and sleep disturbances, based on traditional use only. (Hop Strobiles, Valerian root, Gentian root)

Hops strobile (Humulus lupulus L.) Vervain herb (Verbena officinalis L.) Scullcap herb (Scutellaria lateriflora L.) Jamaica Dogwood root bark (Piscidia erythrina L.) Passiflora herb (Passiflora incarnata L.) Oat seed (Avena sativa L.) Valerian root* (Valerian officinalis L.)	For the temporary relief of symptoms associated with stress, tension and irritability, based on traditional use only.
Hops strobiles (Humulus lupulus L.) Passion Flower herb (Passiflora incarnata L.) Valerian root (Valeriana officinalis L.)	For the temporary relief of symptoms of mild anxiety and to aid sleep, based on traditional use only.
Hops strobiles (Humulus lupulus L.) Passion Flower herb (Passiflora incarnata, L.) Motherwort herb (Leonurus cardiaca L.) Valerian root (Valeriana officinalis L.) Wild Lettuce (Lactuca virosa L.)	A traditional herbal medicinal product used for the temporary relief of symptoms associated with stress such as mild anxiety and for the temporary relief of sleep disturbances, based on traditional use only.
Hops strobiles (Humulus lupulus L.) Scullcap herb (Scutellaria lateriflora L.). Valerian root (Valeriana officinalis L.) Verbena herb (Verbena officinalis L.)	For the temporary relief of symptoms associated with stress such as mild anxiety, based on traditional use only.
Hops strobiles (Humulus lupulus L.) Wild Lettuce leaf (Lactuca virosa L.) Passion Flower herb (Passiflora incarnata L.) Valerian root (Valeriana officinalis L.)	For the temporary relief of sleep disturbances, based on traditional use only.
Hops strobiles (Humulus lupulus L.), Scullcap herb (Scutellaria lateriflora L.), Valerian root (Valeriana officinalis L.)	For the temporary relief of symptoms of mild anxiety, based on traditional use only.
Hops strobiles (Humulus lupulus L.), Scullcap herb (Scutellaria lateriflora L.), Valerian root(Valeriana officinalis L.), Dandelion root (Taraxacum officinale Weber ex Wigg)	For the temporary relief of symptoms associated with stress such as mild anxiety, based on traditional use only.
Horse Chestnut Seed (Aesculus hippocastanum L.)	For the relief of symptoms associated with chronic venous insufficiency and varicose veins, such as tired heavy legs, pain, cramps and swelling

Horse chestnut seeds (Aesculus hippocastanum L. semen)	For the relief of symptoms associated with minor venous insufficiency and varicose veins, such as tired heavy legs, pain, cramps and swelling. This is based on traditional use only.
Horse-chestnut seed (Aesculus hippocastanum L). Calendula flower (Calendula officinalis L.) Witch Hazel bark & root Bark (Hamamelis virginiana L). Peony root (Paeonia officinalis L.)	To relieve the itching and burning associated with haemorrhoids, based on traditional use only.
Horsetail herb	To increase the amount of urine for the purpose of flushing the urinary tract to assist in minor urinary complaints. This is based on traditional use only.
Horsetail (Equisetum arvense L.) herb. Dandelion (Taraxacum officinale Weber) herb and root	To relieve bloating associated with premenstrual water retention, based on traditional use only.
Hypericum perforatum tincture (Hypericum perforatum L. herb) Calendula officinalis tincture (Calendula officinalis L. herb)	For the relief of sore minor cuts and wounds, based on traditional use only.
Ipecacuanha root (Cephaelis ipecacuanha (Brot.) A. Rich.) Liquorice root (Glycyrrhiza glabra L.) Indian Squill bulb (Drimia indica (Roxb.) J P Jessop	For relief of the symptoms of mucus coughs and colds, based on traditional use only.
Ivy (Hedera helix L.) herb Thyme (Thymus vulgaris L.) aerial parts Liquorice (Glycyrrhiza glabra L.) root	For the relief of chesty coughs, mucus coughs and catarrh, based on traditional use only.
Ivy (Hedera helix L.) leaf	To relieve chesty coughs associated with the common cold based on traditional use only.
Ivy Leaf (Hedera helix L.)	To relieve chesty coughs associated with the common cold based on traditional use only
Jamaica Dogwood (Piscidia erythrina L.) Wild Lettuce (Lactuca virosa L.) leaf Passion Flower (Passiflora incarnata L.) herb Hop (Humulus lupulus L.) stobile	For the short term relief of minor aches, tenseness and irritability, based on traditional use only.

Kelp thallus (Ascophyllum nodosum L. (Le Jolis) or Ascophyllum canaliculatum L. (Kuntze)	For the relief of rheumatic or muscular pain, and general aches and pains in the muscles and joints, based on traditionally use only.
Lavender oil (lavandula angustifolia miller, aetheroleum)	For the temporary relief of the symptoms of mild anxiety such as stress and nervousness based on traditional use only.
Lemon Balm Leaf (Melissa officinalis L.)	For the temporary relief of symptoms of mild anxiety, to aid sleep and for mild digestive complaints, such as bloating and flatulence, based on traditional use only
Lemon Balm leaf (Melissa officinalis L.)	For the temporary relief of symptoms of mild anxiety, to aid sleep and for mild digestive complaints, such as bloating and flatulence,based on traditional use only.
Lemon Balm Leaf (Melissa officinalis L.), Valerian Root (Valeriana officinalis L.), Hops Strobiles (Humulus lupulus L.)	For the temporary relief of sleep disturbances due to symptoms of mild anxiety, based on traditional use only.
Lemon Balm Leaf (Melissa officinalis L.), Valerian Root (Valeriana officinalis L.), Passion Flower Herb (Passiflora incarnata L.)	For the Temporary relief ofmild anxiety, to aid sleep and fro mild digestive complaints, such as bloating and flatulence, based on traditional use only
Linseed, whole (Linum usitatissimum L., semen). Senna leaf (Cassia senna L. and/or Cassia angustifolia VAHL, folium). Frangula bark (Rhamnus frangula L., cortex)	For the short term relief of occasional constipation based on traditional use only.
Liquorice Root (Glycyrrhiza glabra L.)	For the symptomatic relief of coughs based on traditional use only.

Lobelia (Lobelia inflata L.) herb. Squill (Urginea maritima L.) bulb	For the relief of coughs, blocked sinuses and catarrh, based on traditional use only.
Lobelia Herb (Lobelia inflata L.) Ipecacuanha Root (Cephaelis Ipecacuanha (Brot.)) White Horehound Herb (Marrubium vulgare L.) Elecampane Root (Inula helenium L.) Hyssop Herb (Hyssopus officinalis L.)	For the symptomatic relief of coughs, based on traditional use only.
Lobelia herb (Lobelia inflata L.) White Horehound herb (Marrubium vulgare L.) Senega Root (Polygala senega L.) Indian Squill bulb (Drimia indica (Roxb.) JP Jessop)	For the symptomatic relief of chesty coughs and catarrh of the throat and chest, based on traditional only.
Lobelia Herb (Lobelia inflata L.), Squill Bulb (Urginea maritima L.)	For the relief of coughs, such as chesty coughs and dry, tickly coughs, based on traditional use only.
Marshmallow Root (Althea officinalis L.) Slippery Elm Bark (Ulmus fulva L.) Sulphur Sublimed Zinc Oxide	For the symptomatic relief of minor skin conditions causing skin irritation such as dermatitis and mild eczema, based on traditional use only.
Matricaria (Matricaria recutita L.) flower	For the symptomatic relief of teething pain and the symptoms associated with teething which are sore and tender gums, flushed cheeks and dribbling based on traditional use only.
Meadowsweet Herb (Filipendula ulmaria (L.) Maxim) Gentian Root (Gentiana lutea L.) Euonymus Bark (Euonymus atropurpureus Jacq.)	For the symptomatic relief of indigestion, heartburn and flatulence, based on traditional use only.
Melissa (Melissa officinalis L.) leaf Valerian (Valeriana officinalis L.) root Passion flower (Passiflora ncarnate L.) herb	For the temporary relief of symptoms of mild anxiety, to aid sleep and for mild digestive complaints, such as bloating and flatulence, based on traditional use only.

Milk Thistle (Silybum marianum (L.) Gaertner) fruits	To relieve the symptoms associated with occasional over indulgence of food and drink such as indigestion and upset stomach based on traditional use only.
Milk Thistle (Silybum marianum (L.) Gaertner) fruits	To relieve the symptoms associated with occasional over indulgence of food and drink such as indigestion and upset stomach based on traditional use only.
Milk Thistle (Silybum marianum (L.) Gaertner) fruits.	To relieve the symptoms associated with occasional over indulgence of food and drink such as indigestion and upset stomach. This is based on traditional use only.
Milk thistle dry extract, milk thistle fruits (silybum marianum (L.) Gaertner)	For the relief of the symptoms associated with occasional over indulgence of food and drink such as indigestion and upset stomach based on traditional use only.
Milk Thistle Fruit (Silybum marianum (L.) Gaertner)	To relieve the symptoms associated with occasional over indulgence of drink and food such as indigestion and upset stomach. This usage is based on traditional use only
Milk Thistle Fruit (Silybum marianum (L.) Gaertner)	To relieve the symptoms associated with occasional over indulgence of food and drink such as headache and upset stomach based on traditional use only.
Milk Thistle Fruit (Silybum marianum (L.) Gaertner)	To relieve the symptoms associated with occasional over indulgence of food and drink such as indigestion and upset stomach based on traditional use only.

Milk Thistle fruits (Silybum marianum (L.) Gaertner)	For the relief of symptoms associated with occasional over indulgence of food and drink such as indigestion and upset stomach.
Nettle (Urtica dioica L.) leaf	for the relief of minor aches and pains in the joints. This is based on traditional use only.
Nettle Root (Urtica dioica L. and/or Urtica urens L.)	To relieve the symptoms of urinary tract discomfort inmen who have been told that they have an enlarged prostate (benign prostatic hyperplasia or BPH), based on traditional use only.
Nettle whole plant (Urtica urens L.) Calendula flower (Calendula officinalis L.) (1:10) Echinacea whole plant (Echinacea angustifolia DC.) St. John's Wort herb (Hypericum perforatum L.)	For the relief of minor burns and scalds, based on traditionally use only.
Oat seed* (Avena sativa L.) Peppermint leaf * (Mentha piperita L.) Lemon balm leaf * (Melissa officinalis L.) Southern Prickly Ash berries (Zanthoxylum clava-herculis L.)	For the symptomatic relief of indigestion, heartburn and stomach ache, based on traditional use only.
Paeonia lactiflora tincture (paeonia lactiflora pallas)	For the symptomatic relief of hot flushes associated with the menopause, based on traditional use only.
Passiflora herb (Passiflora incarnata L.) Wild Lettuce leaf (Lactuca virosa L.) Jamaica Dogwood bark (Piscidia erythrina L.) White Willow bark (Salix alba L.)	For the temporary relief of sleep disturbances, based on traditional use only.
Passion Flower Herb (Passiflora incarnata L.)	For the temporary relief of symptoms associated with stress such as mild anxiety based on traditional use only.
Passion Flower Herb (Passiflora incarnata L.)	For the temporary relief of symptoms associated with stress such as mild anxiety based on traditional use only

Passion Flower Herb (Passiflora incarnata L.)	For the temporary relief of symptoms associated with stress such as mild anxiety based on traditional use only.
Passion Flower Herb (Passiflora incarnata L.)	For the relief of symptoms of mild anxiety and to aid sleep. (Passion flower)
Passion Flower herb (Passiflora incarnata L.) Jamaica Dogwood bark (Piscidia erythrina L.) Hop strobile (Humulus lupulus L.) Valerian root (Valeriana officinalis L.)	To aid sleep, based on traditional use only.
Passion Flower Herb (Passiflora incarnate L.)	For the temporary relief of mild anxiety and to aid sleep, based on traditional use only.
Pelargonium Root (Pelargonium sidoides DC)	For the temporary relief of symptoms of mild anxiety and to aid sleep, based on traditional use only.
Pelargonium Root (Pelargonium sidoidesDC)	For the relief of symptoms of upper respiratory tract infections including common cold, such as sore throat, cough and blocked or runny nose.
Pelargonium Root (Pelargonium sidoidesDC)	For the relief of symptoms of upper respiratory tract infections including common cold, such as sore throat, cough and blocked or runny nose.
Pelargonium Root (Pelargonium sidoidesDC)	To relieve the symptoms of upper respiratory tract infections including the common cold, such as sore throat, cough and blocked or runny nose, based on traditional use only.
Pelargonium sidoides DC	To relieve the symptoms of upper respiratory tract infections including common cold, such as sore throat, cough and blocked or runny nose, based on traditional use only.

Peony Root (Paeonia lactiflora L.)	For the symptomatic relief of hot flushes associated with the menopause, based on traditional use only. (Paeonia lactiflora)
Peppermint oil (essential oil of mentha × piperita L.), eucalyptus oil (essential oil of eucalyptus globulus labilL., eucalyptus), rosemary oil (essential oil of rosmarinus officinalis L.)	For the symptomatic relief of nasal congestion and coughs due to colds, muscular aches and pains, stiffness, based on traditional use only.
Peppermint Oil (Mentha x piperita L.)	For the symptomatic relief of minor digestive complaints such as dyspepsia, flatulence and stomach cramps, based on traditional use only.
Pilewort herb (Ranunculus ficaria L.) Hamamelis bark (Hamamelis virginiana L.) Zinc Oxide Tea Tree leaf oil (Melaleuca alternifolia (Maiden et Beche.) Cheel)	To relieve the itching and burning associated with haemorrhoids, commonly known as piles, based on traditional use only.
Pleurisy root (Asclepias tuberosa L.) Elecampane root (Inula helenium L.) White Horehound leaves and flowering tops (Marrubium vulgare L.) Skunk cabbage root (Symplocarpus foetidus (L.) Salisb.) Lobelia herb (Lobelia inflata L.)	For the symptomatic relief of coughs, based on traditional use only
Pulsatilla herb (Anemone pulsatilla L.) Black Cohosh rhizomes and root (Cimicifuga racemosa (L.) Nutt.) Valerian root* (Valeriana officinalis L.) Scullcap herb* (Scutellaria lateriflora L.) Motherwort flowering herb (Leonurus cardiaca L.)	For the symptomatic relief of menopausal symptoms, such as hot flushes, night sweats, poor sleep, mood changes and irritability, based on traditional use only.
Pumpkin (Cucurbita pepo L.) seeds	For the relief of lower urinary tract symptoms such as frequent urge to urinate and/or increased urinary frequency and urgency in women with overactive bladder, based on traditional use only.

Pumpkin (Cucurbita pepo L.) seeds	For the relief of lower urinary tract symptoms in men who have a confirmed diagnosis of enlarged prostate (benign prostatic hyperplasia; BPH), based on traditional use only. Prior to treatment other serious conditions should have been ruled out by a doctor.
Pumpkin seed (cucurbita pepo L. Convar. Citrullina i. Greb. Var. Styriaca i. Greb), saw palmetto fruit (serenoa repens (bartram) small fructus (sabal serrulata (michaux) nichols fructus)	For the relief of lower urinary tract symptoms in men related to an overactive bladder or bladder weakness, such as urgency to urinate, urinary incontinence, frequent urination based on traditional use only.
Pumpkin seed dry extract (cucurbita pepo L.)	For the relief of lower urinary tract symptoms such as frequent urge to urinate and/or increased urinary frequency and urgency in women with overactive bladder, based on traditional use only.
Pumpkin seed oil (cucurbita pepo L. Convar. Citrullina i. Greb. Var. Styriaca i. Greb)	For the relief of lower urinary tract symptoms related to an overactive bladder or bladder weakness, such as urgency to urinate, urinary incontinence, frequent urination, based on traditional use only.
Pumpkin seed oil (cucurbita pepo L. Convar. Citrullina i. Greb. Var. Styriaca i. Greb), extract of fragrant sumach bark (rhus aromatica aiton)	For the relief of lower urinary tract symptoms in women related to an overactive bladder or bladder weakness, such as urgency to urinate, urinary incontinence, frequent urination based on traditional use only.

Pumpkin seed soft extract (cucurbita pepo L. Convar. Citrullina i. Greb. Var. Styriaca i. Greb)	For the relief of lower urinary tract symptoms in men related to an overactive bladder, such as urgency to urinate and frequent urination, or who have a confirmed diagnosis of benign prostatic hyperplasia (bph), based on traditional use only. Prior to treatment, other serious conditions should have been ruled out by a doctor.
Raspberry Leaf (Rubus idaeus L.ssp. idaeus)	For the symptomatic relief of menstrual cramps, based on traditional use only.
Rhodiola Root (Rhodiola rosea L.)	For the temporary relief of symptoms associated with stress such as fatigue, exhaustion and mild anxiety.
Rhodiola rosea roots and rhizomes	For the temporary relief of symptoms associated with stress, such as fatigue, exhaustion And mild anxiety. This is based on traditional use only.
Rhodiola rosea L roots and rhizomes	For the temporary relief of symptoms associated with stress, such as fatigue, exhaustion and mild anxiety based on traditional use only.
Roman Chamomile flower (Chamaemelum nobile L.)	for the relief of flatulence, bloating and mild upset stomach, based on traditional use only.
Rosemary oil	For the relief of minor muscular and articular pain and minor peripheral circulatory disorders.
Sage (Salvia officinalis L.)	For the relief of excessive sweating associated with menopausal hot flushes, including night sweats based on traditional use only.

Sage Leaf (Salvia officinalis L.)	For the relief of excessive sweating associated with the menopause based on traditional use only.
Saw Palmetto Fruit (Serenoa repens (W. Bart.) Small)	For the relief of lower urinary tract symptoms in men who have a confirmed diagnosis of benign prostatic hypertrophy (BPH)
Saw Palmetto Fruit (Serenoa repens (W. Bart.) Small)	For the relief of lower urinary tract symptoms in men who have a confirmed diagnosis of benign prostatic hypertrophy (BPH), based on traditional use only.
Saw Palmetto Fruit (Serenoa repens (W. Bart.) Small)	To relieve the symptoms of urinary tract discomfort in men who have been told that they have an enlarged prostate (benign prostatic hypertrophy or BPH). This is based on traditional use only.
Scullcap herb (Scutellaria laterifolia L.) Helionas root (Chamaelirium luteum) Jamaican Dogwood bark (Piscidia piscipula) Motherwort herb (Leonurus cardiaca) Buchu leaf (Agathosma betulina (Berg) Pillans)	For the symptomatic relief of menstrual cramps, based on traditional use only.
Senega root (Polygala senega L), Marshmallow root (Althaea officinalis), Ipecacuanha root (Cephaelis ipecacuanha (Brot.) A. Rich.)	For the relief of sore throats and chesty coughs, based on traditional use only.
Senna leaf (Cassia senna L.) Aloes (Cape aloes; Aloe ferox Miller and its hybrids), Cascara bark (Rhamus purshianus D.C.)	For the short term relief of occasional constipation and bloating based on traditional use only.
Senna leaf (Cassia senna L.) Cape Aloe leaf (Aloe ferox Miller) Cascara bark (Rhamnus purshiana D.C.) Dandelion root (Taraxacum officinale Weber ex Wigg.) Valerian root (Valeriana officinalis L.) Myrrh resin (Commiphora myrrha EngL.) Holy Thistle herb (Cnicus benedictus L.)	For the short term relief of occasional constipation, based on traditional use only.

Senna leaf (Cassia senna L.) Cape Aloe leaf(Aloe ferox Miller) Cascara Bark (Rhamnus purshiana D.C.) Dandelion Root (Taraxacum officinale Weber ex Wigg.	For the short term relief of occasional constipation, based on traditional use only.
Senna Leaf Tinnevelly (Cassia angustifolia Vahl). Cape Aloes (Aloe ferox Mill). Cascara Bark (Rhamnus purshianus D.C.). Dandelion Root (Taraxacum officinale F.H. Wigg). Fennel Fruit (Foeniculum vulgare var. dulce Miller)	For the short-term relief of occasional constipation and bloating, based on traditional use only.
Shepherd's Purse herb (Capsella bursa-pastoris L.) Wild Carrot root (Daucus carota L.) Uva Ursi leaf (Arctostaphylos uva-ursi Clivers herb (Galium aparine L.)	A traditional herbal medicinal product used for the symptomatic relief of lower backache including lumbago and sciatica, based on traditional use only.
Sigesbeckia orientalis	For the relief of backache, minor sports injuries, rheumatic or muscular pains and general aches and pains in the muscle and joints, based on traditional use only.
Sigesbeckia orientalis L. subsp. pubescens aerial parts	For the relief of backache, minor sports injuries, rheumatic or muscular pains and general aches and pains in the muscle and joints
Slippery Elm Bark Powder (Ulmus rubra MuhL.)	To relieve sore throats associated with coughs and colds, based on traditional use only.
Spruce Shoots (Picea abies (L.) Karsten)	For the relief of coughs, such as chesty coughs and dry, tickly, irritating coughs, and catarrh based on traditional use only. (Spruce (Piceae abietis)
Squill bulb (Drimia maritima (L.) Stearn.) Capsicum Oleoresin (Capsicum annuum L. var. minimum (Miller) Heiser)	for the relief of coughs, colds and sore throats, based on traditional use only.

Squill Oxymel (Urginea maritima L.)	For the relief of coughs, such as chesty coughs and cough with catarrh, based on traditional use only.
St John's Wort (Hypericum perforatum L.)	To relieve the symptoms of slightly low mood and mild anxiety, based on traditional use only.
St John's Wort aerial parts (Hypericum perforatum L.)	To relieve the symptoms of slightly low mood and mild anxiety based on traditional use only.
St John's Wort Herb (Hypericum perforatum L.)	To relieve the symptoms of slightly low mood and mild anxiety based on traditional use only.
St John's Wort Herb (Hypericum perforatum L.)	For the relief of symptoms of the menopause, including hot flushes, night sweats, slightly low mood and mild anxiety, based on traditional use only.
St John's Wort Herb (Hypericum perforatum L.) & Black Cohosh rhizome and root (Cimicifuga racemosa (L.) Nutt.).	For the relief of symptoms of the menopause, including hot flushes, night sweats, slightly low mood and mild anxiety, based on traditional use only.
St John's Wort Herb (Hypericum perforatum L.) & Black Cohosh rhizome and root (Cimicifuga racemosa (L.) Nutt.).	For the relief of symptoms of the menopause, including hot flushes, night sweats, slightly low mood and mild anxiety, based on traditional use only.
St John's Wort Herb (Hypericum perforatum L.) Black Cohosh rhizome and root (Cimicifuga racemosa (L.) Nutt.).	For the relief of symptoms of the menopause, including hot flushes, night sweats, slightly low mood and mild anxiety based on traditional use only.

St John's Wort Herb (Hypericum perforatum L.), Valerian Root (Valeriana officinalis) & Passion Flower Herb (Passiflora incarnata L.)	To relieve slightly low mood and mild anxiety and sleep disturbances due to symptoms of mild anxiety.
St John's Wort Herb Hypericum perforatum L.)	For the relief of symptoms of slightly low mood and mild anxiety.
St Johns Wort dry extract (Hypericum perforatum L.)	For the relief of slightly low mood and mild anxiety based on traditional use only.
St. John's Wort herb (Hypericum perforatum L.) Calendula flower (Calendula officinalis L.)	for the relief of minor cuts and wounds, based on traditional use only.
St. John's Wort Whole Plant (Hypericum perforatum L.) Yellow Dock Root (Rumex crispus L.) Echinacea angustifolia Whole Plant (Echinacea angustifolia DC.) Labrador Tea Whole Plant (Ledum palustre L), Calendula flower (Calendula officinalis L.)Arnica Whole Plant (Arnica montana L.) ,Pyrethrum flower (Chrysanthemum cinerariaefolium (Trev) Vis.)	For the symptomatic relief of insect bites and stings, based on traditional use only.
Tea tree oil	A traditional herbal medicinal product used for the treatment of small superficial wounds, insect bites, small boils (furuncles and mild acne), for the symptomatic treatment of minor inflammation of the oral mucosa and for the relief of itching and irritation in cases of mild athlete's foot, based on traditional use only
Tea Tree Oil (Melaleuca alternifolia (Maiden and Betche) Cheel) Zinc Oxide Sublimed Sulphur	For the symptomatic relief of minor skin conditions such as spots, pimples, and blemishes, based on traditional use only.
Thyme herb (thymus vulgaris L. and Thymus zygis L., herb)	For the relief of coughs, such as chesty coughs and dry, tickly, irritating coughs and catarrh based on traditional use only.

Thyme herb dry extract (thymus vulgaris L. And thymus zygis L., herb)	For the relief of coughs, such as chesty coughs and dry, tickly, irritating coughs and catarrh, based on traditional use only.
Tinnevelly Senna Leaf (Cassia angustifolia Vahl), Frangula Bark (Rhamnus frangula L.)	For the short term relief of occasional constipation, based on traditional use only.
Uva Ursi leaf (Arctostaphylos uva-ursi (L.) Spreng) Buchu leaf (Agathosma betulina (Berg.)) Clivers herb (Galium aparine L.) Clivers herb) Couchgrass rhizome (Agropyron repens Beauvais) Horesetail herb (Equisetum arvense L.) Shepherd's Purse herb (Capsella bursa-pastoris L.)	To help flushing of the urinary tract and to assist in minor urinary complaints associated with cystitis in women, based on traditional use only.
Uva Ursi leaf (Arctostaphylos uva-ursi (L.). Spreng) Buchu Leaf Powder (Agathosma betulina. (Berg.) Pillans) Parsley Piert herb (Aphanes arvensis L.)	A traditional herbal medicinal product used to relieve symptoms in mild cases of water retention, based on traditional use only.
Uva ursi leaf (Arctostaphylos uva-ursi; (L.) Spreng., folium), Clivers herb; (Galium aparine L. herba), Burdock, root (Arctium lappa L. radix)	To relieve symptoms in mild cases of water retention, based on traditional use only.
Valerian root (Valerian officinalis) Hops Strobiles (Humulus lupulus L.)	
Valerian root (Valeriana officinalis L)	To aid sleep and for the temporary relief of sleep disturbances based on traditional use only.
Valerian root (Valeriana officinalis L)	For the temporary relief of sleep disturbances due to symptoms of mild anxiety based on traditional use only.
Valerian root (Valeriana officinalis L)	For the temporary relief of sleep disturbances and mild anxiety based on traditional use only
Valerian root (Valeriana officinalis L)	For the temporary relief of symptoms of mild anxiety and to aid sleep, based on traditional use only.

Valerian root (*Valeriana officinalis L, radix*), Hop strobile (*Humulus lupulus L.*), Scullcap herb (*Scutellaria lateriflora L.*)	For the temporary relief of symptoms associated with stress such as mild anxiety, and to aid sleep, based on traditional use only.
Valerian Root (*Valeriana officinalis L.*)	For the temporary relief of sleep disturbances due to symptoms of mild anxiety based on traditional use only.
Valerian Root (*Valeriana officinalis L.*)	For the temporary relief of sleep disturbances based on traditional use only
Valerian Root (*Valeriana officinalis L.*)	To relieve slightly low mood, mild anxiety and sleep disturbances due to mild anxiety, based on traditional use only.
Valerian root (*Valeriana officinalis L.*)	For the temporary relief of symptoms associated with stress such as mild anxiety and to aid sleep, based on traditional use only.
Valerian root (*Valeriana officinalis L.*) Fennel fruit (*Foeniculum vulgare Miller subsp. vulgare var. dulce (Miller) Thellung*) Myrrh gum-resin (*Commiphora molmol Engler*)	To relieve symptoms of dyspepsia and flatulence, based on traditional use only.
Valerian root (*Valeriana officinalis L.*) Fennel fruit (*Foeniculum vulgare Miller subsp. vulgare var. dulce (Miller) Thellung*) Myrrh gum-resin (*Commiphora molmol Engler*	To relieve symptoms of dyspepsia, based on traditional use only.
Valerian Root (*Valeriana officinalis L.*) St. John's Wort Herb (*Hypericum perforatum L.*)	To relieve slightly low mood, mild anxiety and sleep disturbances due to mild anxiety, based on traditional use only.
Valerian Root (*Valeriana officinalis L.*)	For the temporary relief of symptoms of mild anxiety and to aid sleep.

Valerian Root (Valeriana officinalis L.)	For the temporary relief of sleep disturbances due to symptoms of mild anxiety based on traditional use only.
Valerian Root (Valeriana officinalis L.), Hops	For the temporary relief of sleep disturbances caused by the symptoms of mild anxiety based on traditional use only.
Valerian Root (Valeriana officinalis L.), Hops Strobiles (Humulus lupulus L.)	For the temporary relief of symptoms associated with stress such as mild anxiety, based on traditional use only
Valerian Root (Valeriana officinalis L.), Hops Strobiles (Humulus lupulus L.) Lemon Balm Leaf (Melissa officinalis L.)	For the temporary relief of sleep disturbances due to symptoms of mild anxiety based on traditional use only.
Valerian Root (Valeriana officinalis L.), Hops Strobiles (Humulus lupulus L.) Lemon Balm Leaf (Melissa officinalis L.)	For the temporary relief of symptoms of mild anxiety, based on traditional use only. (Valerian, Lemon Balm)
Valerian Root (Valeriana officinalis L.), Passion Flower Herb (Passiflora incarnata L.)	For the temporary relief of sleep disturbances due to symptoms of mild anxiety based on traditional use only.
Valerian root (Valeriana officinalis L.), Passion flower herb (Passiflora incarnata L.), Wild lettuce leaf (Lactuca virosa L.)	To aid sleep based on traditional use only.
Valerian root (Valeriana officinalis L.), Verbena herb (Verbena officinalis L.), Gentian root (Gentiana lutea L.), Scullcap herb (Scutellaria lateriflora L.)	For the temporary relief of sleep disturbances and mild anxiety, based on traditional use only.
Valerian Root (Valeriana officinalis) & Passion Flower Herb (Passiflora incarnata L.)	For the temporary relief of sleep disturbances due to symptoms of mild anxiety

Valerian Root (Valeriana officinalis) (Scullcap (Scutellaria lateriflora L.)	For the short term relief of symptoms associated with stress such as mild anxiety, tenseness or irritability, based on traditional use only.
Valerian root dry extract (valeriana officinalis L.)	For the temporary relief of sleep disturbances based on traditional use only.
Valerian root tincture (valeriana officinalis L.)	For the temporary relief of symptoms of mild anxiety and to aid sleep, based on traditional use only.
Valeriana officinalis L.	For the temporary relief of sleep disturbances and mild anxiety based on traditional use only.
Valerianroot (Valeriana officinalis L.) Hops strobile (Humulus lupulus L.)	to aid sleep based on traditional use only
Verbena Herb (Verbena officinalis L.) Hops Strobiles (Humulus lupulus L.), Valerian Root (Valeriana officinalis L.), Passion Flower Herb (Passiflora incarnata L.), Wild Lettuce (Lactuca virosa L.)	For the temporary relief of sleep disturbances based on traditional use only. (Verbena, hops, valerian, passion flower and Wild Lettace)
Verbena Herb (Verbena officinalis L.), Hop strobiles (Humulus lupulus L.), Wild Lettuce leaf (Lactuca virosa L.), Passion Flower herb (Passiflora incarnata L.)	For the temporary relief of sleep disturbances, based on traditional use only.
Vervain (Verbena officinalis L.) herb Scullcap (Scutellaria lateriflora L.) herb Passion Flower (Passiflora incarnata L.) herb. Oat (Avena sativa L.) seed	For the temporary relief of symptoms associated with stress, such as mild anxiety and to aid sleep, based on traditional use only.
White horehound herb (Marrubium vulgare L.) Sage leaf (Salvia officinalis L.) Yarrow herb (Achillea millefoliumL.) Vervain herb (Verbena officinalis L.) Lobelia herb (Lobelia inflata L.)	To relieve symptoms of hayfever, based on traditional use only.

White Willow Bark (Salix alba L.) Black Cohosh Root (Cimicifuga racemosa Nutt.) Scullcap herb (Scutellaria laterifolia L.) Yarrow Herb (Achilliea millefolium L.) Burdock Root (Articum lappa L.)	For the symptomatic relief of backache, rheumatic or muscular pain, and general aches and pains in the muscles and joints based on traditional use only.
Wild Carrot herb (Daucus carota L.) Pellitory herb (Anacyclus pyrethrum L.) Buchu leaf (Agathosma betulina Juniper berry(Juniperus communis L.) Clivers herb (Galium aparine L.)	To relieve symptoms in mild cases of water retention, based on traditional use only.
Wild indigo root (Baptisia tinctoria (L.) R.Br.) Echinacea purpurea root (Echinacea purpurea (L.) Moench), Echinacea pallida root (Echinacea pallida (Nutt.) Nutt.), White cedar tips and leaves (Thuja occidentalis L.)	To relieve the symptoms of the common cold, such as cough, catarrh, sore throat, runny or blocked nose, based on traditional use only.
Willow bark (Salix alba L.) Passion Flower herb (Passiflora incarnata L.) Valerian root (Valeriana officinalis L.)	For the relief of backache, rheumatic pain and general aches and pains in the muscles and joints, based on traditional use only.
Witch Hazel bark (Hamamelis virginiana L.)	To relieve the itching and burning associated with haemorrhoids, based on traditional use only.
Yarrow Herb (Achillea millefolium L.) Elder Flower (Sambucus nigra L.) Prickly Ash, Southern Bark (Zanthoxylum clavaherculis L.) Uva Ursi Leaf (Arctostaphylo uva-ursi (L.) Spreng) Burdock Root (Arctium lappa L.) Clivers Herb (Galium aparine L.) Poplar Bark (Populus tremuloides Michx.)	For the symptomatic relief of rheumatic pain and general aches and pains in the muscles and joints, based on traditional use only.
Yarrow herb (Achillea millefolium L.) Vervain herb (Verbena officinalisL.) White Horehoundherb(Marrubium vulgare L.) Sage leaf (Salvia officinalisL.)	To relieve the symptoms of nasal catarrh and catarrh of the throat, based on traditional use only.

Appendix 2.

MHRA Guidance on Banned and Restricted Herbal Ingredients

MHRA Guidance on Banned and restricted herbal ingredients – December 2014

Common name	Botanical source
Aconite	All *Aconitum* species
Adonis vernalis	*Adonis vernalis*
African serpentwood	*Rauwolfia vomitoria*
Areca	*Areca catechu*
Belladonna herb	*Atropa belladonna - herb, Atropa acuminata - herb*
Belladonna root	*Atropa belladonna - root, Atropa acuminata - root*
Canadian Hemp	*Apocynum, cannabinium*
Catha	*Catha edulis*
Celandine	*Chelidonium majus*
Chenopodium	*Chenopodium ambrosioides var anthelminticum*
Cinchona bark	*Cinchona calisaya, Cinchona ledgerana, Cinchona officinalis, Cinchona succirubra, Cinchona micrantha*
Colchicum corm	*Colchicum autumnale*
Conium leaf and fruits	*Conium maculatum*
Convallaria	*Convallaria majalis*
Crotalaria fulva	*Crotalaria berberoana*
Crotalaria spect	*Crotalaria spectabilis*
Cucurbita	*Curcurbita maxima*
Duboisia	*Duboisia myoporoides, Duboisia leichardtii*
Elaterium	*Ecballium elaterium*
Embelia	*Embelia ribes, Embelia robusta*

Ephedra	*Ephedra sinica, Ephedra equisetina, Ephedra distachya, Ephedra intermedia, Ephedra gerardiana*
Ergot, prepared Ergot of Rye; Smut of Rye; Spurred Rye;	*Claviceps purpurea*
Erysimum	*Erysimum canescens*
Fangji	*Cocculus laurifolius, Cocculus orbiculatus, Cocculus trilobus*
Fangji	*Stephania tetrandra*
Foxglove	*Digitalis leaf, Digitalis prepared*
Gelsemium	*Gelsemium sempervirens*
Holarrhena	*Holarrhena antidysenterica*
Hyoscyamus	*Hyoscyamus niger, Hyoscyamus albus, Hyoscyamus muticus*
Ignatius bean	*Strychnos ignatii; S. cuspida*
Indian Poke	*Veratrum viride, Hellebore American; Green Hellebore; American Veratrum;*
Jaborandi	*Pilocarpus jaborandi, Pilocarpus microphyllus*
Kamala	*Mallotus philippinensis*
Kava-kava	*Piper methysticum*
Kousso	*Brayera anthelmintica*
Lobelia	*Lobelia inflata*
Male fern	*Dryopteris filix-mas*
Mandrake	*Mandragora autumnalis*
May Apple; Devil's Apple; Wild Lemon; Indian podophyllum	*Podophyllum*
May Apple; Devil's Apple; Wild Lemon; Indian podophyllum	*Podophyllum indian*

May Apple; Devil's Apple; Wild Lemon; Indian podophyllum	*Podophyllum resin*
Mistletoe berry	*Viscum album*
Mu tong	*Clematis armandii, Clematis montana*
Mu tong; Fangji; Birthwort; Long Birthwort; Indian Birthwort	*Aristolochia, Aristolochia clematis, Aristolochia contorta, Aristolochia debelis, Aristolochia fang-chi, Aristolochia manshuriensis, Aristolochia serpentaria*
Mu tong; Fangji; Birthwort; Long Birthwort; Indian Birthwort	*Aristolochia, Aristolochia clematis, Aristolochia contorta, Aristolochia debelis, Aristolochia fang-chi, Aristolochia manshuriensis, Aristolochia serpentaria*
Mu tong	*Akebia quinata, Akebia trifoliata*
Poison Ivy	*Rhus radicans*
Poison Nut	*Strychnos nux vomica seed, Nux vomica seed;*
Poison Oak	*Rhus toxicodendron*
Pomegranate Bark	*Punica granatum*
Poppy capsule	*Papaver somniterum*
Quebracho	*Aspidosperma quebrachoblanco*
Ragwort; Groundsel	*Senecio*
Rauwolfia; serpentwood; Indian snakeroot	*Rauwolfia serpentina*
Sabadilla; Cevadilla	*Schoenocaulon officinale*
Santonica	*Artemisia cina*
Savin	*Juniperus sabina*
Scopolia	*Scopolia carniolica, Scopolia japonica*
Slippery Elm Bark (whole or unpowdered)	*Ulmus fulva, Ulmus rubra*
Stavesacre seeds	*Delphinium staphisagria*

Stramonium	*Datura stramonium, Datura innoxia*
Strophanthus	*Strophanthus kombe, Strophanthus courmonti, Strophanthus nicholsoni, Strophanthus gratus, Strophanthus emini, Strophanthus sarmentosus, Strophanthus hispidus*
White Hellebore	*Veratrum album*
Yohimbe bark	*Pausinystalia yohimbe*
Indian berry	*Cocculus indicus*

References

Glossary of terms

Bibliography

Further Reading

Glossary of Terms

Abortifacient – a medicine for inducing abortion cf *emmenagogue*

Adjuvant – an agent administered during or after chemotherapy to prevent secondary tumor formation.

ADMET - absorption, distribution, metabolism, excretion, and toxicology.

Analog – a compound that is either functionally or structurally similar to another.

Anion - an ion in which total number of electrons is more than the total number of protons, giving it a negative charge.

Antiemetic – a drug for preventing vomiting.

Anthelminthic/antihelmintic – a drug for expelling worms or parasites.

Antihypertensive – a drug for reducing blood pressure.

Antioxidant - an agent which prevents oxidation (loss of an electron) within an organism.

Anxiolytic – a drug used to lessen or prevent anxiety.

Aromatic – a cyclic (ring-shaped), flat molecule with a ring of bonds that exhibits more stability than other structural arrangements containing the same set of atoms.

Astringent – an agent which causes skin tightening and contraction.

Authority - in botany this refers to the individual with whom a species' discovery is linked - e.g. *Genus species* followed by L. for Linnaeus.

Benefit-sharing – a legal or moral framework for ensuring that benefits derived from the commercialisation of genetic resources or knowledge are eqitably distributed.

Binomial - a two part name, commonly in the format *Plant species* followed by botanical authority.

Bioavailability - the proportion of a drug or other substance which enters the circulation when introduced into the body and is able to exert activity.

Biocide - a chemical substance or micro-organism which can destroy or render harmless an otherwise harmful organism.

Blockbuster drug – a drug that generates annual sales of at least $1 billion US dollars.

Carminative – a drug used for the purpose of inducing a calm or relaxed state.

Cation - an ion in which total number of electrons is less than the total number of protons, giving it a positive charge.

Chemotaxonomy - the classification of chemical characteristics and the study their occurence in different plant families or taxa.

Chromatogram – the visual output from a chromatographic device.

Cyclotide – a type of peptide produced by plants with anti-microbial and anti-viral effects.

Cytochrome P450 – a type of enzyme which facilitates the break down and processing of many drug products in the liver.

Derivatized – transformation of a chemical compound into a similar compound, or derivative.

Diaphoretic – a drug to induce perspiration.

Doctrine of Signatures – a belief based on an association between the appearance, smell, taste or habitat of a plant and its likely medicinal purpose.

Emetic – a drug for inducing vomiting.

Enzyme - a protein which catalyses (speeds up) the conversion of one substance to another.

Ethnobotany - the socio-cultural and anthropological study of the relationship between plants and the people who use them.

Ethnopharmacology - the scientific study of the relationship between traditional medicines and the people who use them.

Excipient – non-active ingredients added to a medicinal product.

Exudate - a rich protein squeezed from its source.

Febrifuge – a drug for reducing a fever.

Generic drug - copies of brand-name drugs that have exactly the same pharmacological properties

Humoral - a name given to medical systems which view health as a complex interplay between humoral elements - e.g. bile, phlegm, wind or fire.

Hypotensive – a drug for lowering blood pressure.

Immunosupressant – a substance which reduces or prevents normal immune response.

Ion – an atom in which the total number of electrons is not equal to its total number of protons; Ionized/ionization – the process through which an atom gains or loses electrical charge.

In vitro - from the Latin for 'in glass', *in vitro* refers to testing which takes place in test tubes or microplates.

In vivo - Latin for 'in the living', *in vivo* describes tests which occur in microorganisms, animals and humans.

Isotope – a variant of an element containing a different number of neutrons, e.g. ^{13}C is an isotope of ^{12}C.

Linnaean - attributed to Carl Linnaeus (1707-1778) who devised a classification system based on the sexual characteristics of plants. Names from his system are often referred to as Latin names.

Magic bullet (*sometimes Silver bullet*) – a term devised by the German scientist Paul Ehrlich taken to mean a perfectly targeted drug without side effects.

Materia Medica - a generic term for the medicines and medical practices of a country or a healing system, derived from Dioscorides' *De Materia medica*.

Mitosis – part of the cell cycle where replicated chromosomes are separated into two new nuclei in eukaryotic (nucleus containing) cells.

Meridian – literally a line or channel, used in TCM to describe vessels which distribute *qi*.

Metabolomics – the study of the metabolome – the collection of all metabolites in a biological cell, tissue, organ or organism, which are the end products of cellular processes.

Mole (mol.) – an amount or sample of a chemical substance containing exactly $6.022,140,76 \times 10^{23}$ elementary entities.

Molar limit – the point at which only one molecule per litre is present in a mixture (1 part in 1×10^{24}).

Molecular weight – the relative weight of one molecule of a substance to ^{12}C (12.000 amu) measured in atomic mass units (amu).

Morphology – the scientific study of shape or form, especially in relation to plants.

Moxibustion – the practice of burning *Artemisia vulgaris* L. (mugwort) leaves on or above the skin to stimulate the flow of *qi*.

Naturopathy – an approach which encourages 'self-healing' through herbal, homeopathic and natural medicine.

Nephrotoxic - poisonous to the kidneys.

Oncogenic - cancer causing.

Orthopaedic Aid – a medicine to help the function of the musculoskeletal system.

Panacea – a drug to be used for the treatment of any condition.

Pathogen – a disease causing agent.

Pharmacognosy – an academic field relating to the study of natural products, their identification, pharmacology and ethnomedical context.

Pharmacopeia – a collection of reference standards for the production and formulation of medicines.

Pharmacovigilance – the monitoring of pharmaceutical products after they have been licensed for use.

Phenolic – an aromatic ring structure with a hydroxy (OH) group attached to at least 1 carbon.

Phytoalexin – a compound produced by a plant to combat parasitic infection.

Placebo – a substance without therapeutic effect, often given as a substitute for medicines in clinical trials. From the Latin *placebo* or 'I shall please'.

Polymorphism - the ability of a solid material to exist in more than one form or crystal structure (known in chemistry as allotropy).

Pulmonary Aid – a medicine to help with lung conditions.

Retention time - the time for which a compound binds to a stationary phase (e.g. a column or silica sheet).

Scurvy – an 'historic' disease caused by diets severely lacking in vitamin C ($C_6H_8O_6$ ascorbic acid).

Spectra – a series of plotted values which demonstrate the sbundance and location of analytes.

Stereochemistry – the spatial properties and atomic arrangement of elements or compounds.

Structure elucidation – the process of interpreting analytical data into chemical structures.

Sublimation – direct transition of a solid substance into a gaseous state without passing through a liquid state.

Supercritical fluid extraction – extraction using changes in temperature and pressure parameters to convert the solvent from a liquid to a gas phase.

Synergism – the enhanced effect resulting from two or more medicinal compounds acting together.

Tetracyclic – a chemical structure consisting of four connected rings.

Tonic – a drug taken to restore or promote perfect health or well-being.

Volatile – compounds that can easily transition into a gaseous state (see also sublimation).

Bibliography

Ahmad, S., 2008. Unani Medicine: Introduction and Present Status in India. Internet Journal of Alternative Medicine, 6(1), p.8.

Ayres, P. 2015. *Britain's Green Allies: Medicinal Plants in Wartime.* Leicester: Troubador (Matador).

Barber, S., 2014. Regulation of herbal medicines - Commons Library Standard Note. UK Parliament. Available at: http://parliament.uk/briefing-papers/SN06002/regulation-of-herbal-medicines.

Booker, A., Agapouda, A., Frommenwiler, D.A., Scotti, F., Reich, E. and Heinrich, M. 2018. St John's wort (Hypericum perforatum) products – an assessment of their authenticity and quality. Phytomedicine. 40, pp. 158-164.

Booker, A., Frommenwiler, D., Reich, E., Horsfield, S. and Heinrich, M. 2016. Adulteration and Poor Quality of Ginkgo biloba Supplements. Journal of Herbal Medicine. 6 (2), pp. 79-87.

Debelle, F.D., Vanherweghem, J.-L. & Nortier, J.L., 2008. Aristolochic acid nephropathy: A worldwide problem. Kidney International, 74, pp.158–169.

European Medicines Agency, 2015. European Medicines Agency pre-authorisation procedural advice for users of the centralised procedure. , 2(April), pp.1–144.

Friedrich, W. 2014. Well-established use applications in EEA Regulatory Affairs. , pp.1–16.

Halushka, P.V. 2009. St. John's Wort: A Mini-Review of its Pharmacokinetics and Anti-Depressant Effects. AccessMedicine from McGraw-Hill

Hancock, T. & Wang, X., 2017. Traditional Chinese medicine seeks clinical legitimacy https://ft.com/content/65018acc-5d79-11e7-b553-e2df1b0c3220

Harvey, A.L., Edrada-Ebel, R. & Quinn, R.J., 2015. The re-emergence of natural products for drug discovery in the genomics era. Nature Reviews Drug Discovery, 14(2), pp.111–129.

Heinrich, M. & Anagnostou, S., 2017. From Pharmacognosia to DNA-Based Medicinal Plant Authentication - Pharmacognosy through the Centuries. *Planta Medica*, 83(14–15), pp.1110–1116.

Heinrich, M. & Scotti, F. 2018. unpublished work on food-medicine interface.

Houghton, P.J. and Raman, A., 1998. *Laboratory Handbook for Fractionation of Natural Extracts*. London: Chapman and Hall,

Jiao, L. et al., 2017. Effects of Chinese medicine as adjunct medication for adjuvant chemotherapy treatments of non-small cell lung cancer patients. Scientific Reports, 7(March).

Hsiao-Yu Yang, Pau-Chung Chen, and Jung-Der Wang, 2014. Chinese Herbs Containing Aristolochic Acid Associated with Renal Failure and Urothelial Carcinoma: A Review from Epidemiologic Observations to Causal Inference. BioMed Research International, vol. 2014.

Jones, A., 2015. *Chemistry: An Introduction for Medical and Health Sciences*. John Wiley & Sons.

Lam, W. et al., 2015. PHY906 (KD018), an adjuvant based on a 1800-year-old Chinese medicinal mixture (Huangqin Tang), enhanced the anti-tumor activity of Sorafenib by changing the tumor microenvironment. Scientific Reports, 5, p.9384.

Low, T. Y., Wong, K. O., Yap, A. L. L., De Haan, L. H. J. and Rietjens, I. M. C. M., 2017. The Regulatory Framework Across International Jurisdictions for Risks Associated with Consumption of Botanical Food Supplements. Comprehensive Reviews in Food Science and Food Safety, 16: 821–834

MHRA, 2012. a Guide To What Is a Medicinal Product. MHRA Guidance Note No. 8 Revised, (8), p.27.

Mukherjee, P.K. and Houghton, P.J., 2009. *Evaluation of Herbal Medicinal Products Perspectives on quality, safety and efficacy.* London: Pharmaceutical Press.

Newman, D.J. & Cragg, G.M., 2016. Natural Products as Sources of New Drugs from 1981 to 2014. Journal of Natural Products, 79(3), pp.629–661.

The Nuffield Council on Bioethics, 2005. Chapter 8: The use of animals for research in the pharmaceutical industry. , pp.133–151.

Ortiz, M. et al., 2017. World Congress Integrative Medicine & Health 2017: part three. BMC Complementary and Alternative Medicine, 17(S1), p.333.

Palhares, R.M. et al., 2015. Medicinal plants recommended by the world health organization: DNA barcode identification associated with chemical analyses guarantees their quality. PLoS ONE, 10(5), pp.1–29.

Porter, R. (ed.), 2006. *The Cambridge History of Medicine.* Cambridge: Cambridge University Press.

Serban, C., Sahebkar, A., Ursoniu, S. et al., 2015. Effect of sour tea (Hibiscus sabdariffa L.) on arterial hypertension: a systematic review and meta-analysis of randomized controlled trials. Journal of Hypertension. Volume 33 - Issue 6 - p 1119–1127.

Sharma, P., Murthy, P. & Bharath, M.M.S., 2012. Chemistry, metabolism, and toxicology of cannabis: Clinical implications. Iranian Journal of Psychiatry, 7(4), pp.149–156.

Shepard, G.H. 2004. A Sensory Ecology of Medicinal Plant Therapy in Two Amazonian Societies. American Anthropologist, 106, 2, pp. 252–266.

Tilton, R., Paiva, A. A., Guan, J.-Q., Marathe, R., Jiang, Z., van Eyndhoven, W., Cheng, Y.-C., 2010. A comprehensive platform for quality control of botanical drugs (PhytomicsQC): a case study of Huangqin Tang (HQT) and PHY906. Chinese Medicine, 5, 30.

Trimble, E.L. & Rajaraman, P., 2017. Integrating Traditional and Allopathic Medicine: An Opportunity to Improve Global Health in Cancer. JNCI Monographs, 2017(52), pp.2017–2018.

Urbanowitz, S. & Bishop, C., 2015. Good Agricultural Practices (GAP) and Good Handling Practices (GHP): Risk Mitigation in Edible Horticultural Production Systems.

Vane JR, Botting RM, 2003. The mechanism of action of aspirin. Thromb Res.; 110:255–258.

World Health Organization (WHO). 2010. Benchmarks for training in traditional /complementary and alternative medicine: Benchmarks for training in Unani medicine. World Health Organization.

World Health Organization (WHO). 2013. WHO Traditional Medicine Strategy 2014-2023. World Health Organization (WHO), pp.1–76.

Further Reading

Badal McCreath, S., Delgoda, R. (2017). *Pharmacognosy: Fundamentals, Applications and Strategies.* London: Elsevier, Academic Press.

Christenhusz, M.J.M., Fay M.F., Chase M.W. (2017). *Plants of the World.* Kew: Kew Publishing,

Hanson, B.A. (2005). *Understanding Medicinal Plants.* The Haworth Herbal Press, New York.

Heinrich, M., Barnes, J., Prieto-Garcia, J.M., Gibbons, S., Williamson, E.M. (2018). *Fundamentals of Pharmacognosy and Phytotherapy.* 3e. London: Elsevier.

Heinrich, M., Jäger, A.K. (2015). *Ethnopharmacology.* Chichester: Wiley-Blackwell

Houk, C.C., Post, R. (1996). *Chemistry: Concepts and Problems: A Self-Teaching Guide. Wiley Self–Teaching Guides.* 2e. New York: Wiley.

Leon, C. and Yu-Lin, L. (2017). *Chinese Medicinal Plants, Herbal Drugs and Substitutes: an identification guide.* Kew: Kew Publishing,

Preston, C.L., (2016). *Stockley's Drug Interactions: A Source Book of Interactions, Their Mechanisms, Clinical Importance and Management.* 11e. London: Pharmaceutical Press.

Simmonds, M.S.J., Howes, M-J, Irving, J. (2017) *The Gardener's Companion to Medicinal Plants.* Kew: Frances Lincoln in association with RBG Kew.

φ About Phytoversity

An Introduction to Natural Medicine: From Plant to Patient is Phytoversity's first publication. Phytoversity aims to put people in touch with medicinal plants by:

- *Providing accurate and accessible information about medicinal plants.*
- *Raising the profile of medicinal plants in a responsible and unbiased way.*
- *Supporting and promoting best practice in sustainability and sourcing of medicinal plants*

To sign up to our mailing list and receive details of special offers, promotions and new titles, please visit: **https://phytoversity.org**

Ordering Further Copies of This Book

Re-orders – to order further physical copies of this book please visit your local book store or Amazon online store at https://amazon.com

E-book edition – an electronic version of this book is also available from May 2018 from the Kindle book store at https://amazon.com

Printed in Great Britain
by Amazon